Girls Into Maths Can Go

Edited by

LEONE BURTON

This book accompanies the Open University/Inner London Education Authority Course 'Girls into Maths' (PM 645)

HOLT, RINEHART AND WINSTON
London · New York · Sydney · Toronto

Holt, Rinehart and Winston Ltd: 1 St Anne's Road,
Eastbourne, East Sussex, BN21 3UN

British Library Cataloguing in Publication Data
Girls into maths can go.
1. Mathematics—Study and teaching—Great Britain 2. Sex discrimination in education—Great Britain
I. Burton, Leone
510'.7'1041 QA14.G7
ISBN: 0-03-910687-X

Typeset by Scribe Design, Gillingham, Kent
Printed in Great Britain by Biddles Ltd, Guildford

Last digit is print no: 9 8 7 6 5 4 3 2 1

Acknowledgements

The publishers are grateful to the following for their kind permission to reproduce copyright material in this book:

Chapter 2 © *Mathematics in Schools* and the Controller of Her Majesty's Stationery Office.

Chapter 3 © the Equal Opportunities Commission, Overseas House, Quay Street, Manchester, M3 3HN.

Chapter 5 © Rosalinde Scott-Hodgetts.

Chapter 7 © of the *British Journal of Sociology of Education.*

Chapter 8 © Northam, J.A.

Chapter 9 © *New Scientist*, New Science Publications.

Chapter 10 © the University of London Institute of Education.

Chapter 11 © the Microelectronics in Education Programme.

Chapters 12 and 15 © *Mathematics Teaching* and the ATM Supplements, published by the Association of Teachers of Mathematics, Kings Chambers, Queen Street, Derby DE1 3DA.

Chapter 16 © *For the Learning of Mathematics* published by FLM Publishing Association, 4436 Marcil Avenue, Montreal, Quebec, Canada, H4A 2Z8.

Chapter 17 © Zelda Isaacson.

Appendix, Women in Mathematics—Herstory, the Women in Mathematics and Science Kit produced by the Participation and Equity Program (PEP), Victoria, Australia. The PEP Committee is an advisory committee to the Victorian Minister of Education.

The editor wishes to record her thanks to the rest of the course team of the in-service pack 'Girls into Maths' for their help and support and to Rita Pryor for her hard work and patience.

Notes on Contributors

Barbara Binns was writing as a mathematics teacher in Stockport and is now a research fellow at the Shell Centre for Mathematical Education.

Stephen Brown is Professor of Philosophy of Education and Mathematics Education at the State University of New York at Buffalo, USA, and has directed his teaching, research and writing towards creating the social, personal and humanistic contexts of mathematics education.

Leone Burton is Head of the Mathematics Division in the Faculty of Education and Community Studies, Thames Polytechnic, London. Her concerns in mathematics education lie in the development and use of mathematical thinking.

Alan Eales is Head of Mathematics at Oadby Beachamp College, Leicester.

Derek Foxman is Head of the Department of Mathematics at the National Foundation for Educational Research in England and Wales.

Mary Gribbin is a teacher and freelance writer specialising in social science.

Zelda Isaacson is Senior Lecturer in Mathematics Education at the Polytechnic of North London, having previously taught mathematics in secondary schools.

Lynn Joffe is a Senior Research Officer with the Department of Mathematics at the National Foundation for Educational Research in England and Wales.

Alison Kelly is a lecturer in the Department of Sociology, University of Manchester.

Gilah C. Leder is a Senior Lecturer in the Faculty of Education at Monash University, Melbourne, Australia.

Jean Northam is Senior Lecturer in Education at Rolle College, Exmouth.

Rosalinde Scott-Hodgetts is a Lecturer in Mathematics Education at the Polytechnic of the South Bank, London.

Hilary Shuard is Deputy Principal of Homerton College, Cambridge and Director of the SCDC Primary Mathematics Project.

Dale Spender is an author of many books on feminist issues in education and is editor of the journal *Women's Studies International Forum*.

Anita Straker was Director of the Microelectronics Education Programme, Primary Project, and is now Inspector of Mathematics in the Inner London Education Authority.

Hazel Taylor is Adviser for Equal Opportunities in the London Borough of Brent.

Ruth Townsend is a Lecturer in the Mathematics Division of the Faculty of Education and Community Studies, Thames Polytechnic.

Rosie Walden was Research Officer in the Girls and Mathematics Unit of the University of London Institute of Education, and now teaches in Hackney, London.

Valerie Walkerdine is Director of the Girls and Mathematics Unit and Lecturer in Curriculum Studies at the University of London Institute of Education.

Robin Ward was EOC Project Director with the Information Technology Unit in the London Borough of Croydon and now teaches Information Technology with the Technical and Vocational Education Initiative in Croydon.

Contents

Acknowledgements v
Notes on Contributors vii
Introduction *Leone Burton* 1

PART I: WHAT IS THE PROBLEM? 21

1 The Relative Attainment of Girls and Boys in Mathematics in the Primary Years 23
Hilary Shuard

2 Attitudes and Sex Differences – Some APU Findings 38
Lynn Joffe and Derek Foxman

3 Sidetracked? A Look at the Careers Advice Given to Three Fifth-Form Girls 51

4 Some Thoughts on the Power of Mathematics 58
Dale Spender

5 Girls and Mathematics: The Negative Implications of Success 61
Rosalinde Scott-Hodgetts

6 Mathematics Learning and Socialisation Processes 77
Gilah C. Leder

7 Gender Roles at Home and School 90
Alison Kelly with Juliet Alexander, Umar Azam, Carol Bretherton, Gillian Burgess, Alice Dorney, Julie Gold, Caroline Leahy, Anne Sharpley and Lin Spandley

8 Girls and Boys in Primary Maths Books 110
Jean Northam

9 Boys Muscle in on the Keyboard 117
Mary Gribbin

10 Characteristics, Views and Relationships in the Classroom 122
Rosie Walden and Valerie Walkerdine

PART II: WHAT CAN BE DONE? 147

11 Should Mary have a little Computer? 149
Anita Straker

12 The Girls in my Tutor Group will not Fail at Maths ... 153
Barbara Binns

13 Experience with a Primary School Implementing an Equal Opportunity Enquiry 156
Hazel Taylor

14 Girls and Mathematics at Oadby Beachamp College 163
Alan Eales

15 Girl-Friendly Mathematics 187
Leone Burton and Ruth Townsend

16 The Logic of Problem Generation: from Morality and Solving to De-posing and Rebellion 196
Stephen I. Brown

17 Freedom and Girls' Education: A Philosophical Discussion with Particular Reference to Mathematics 223
Zelda Isaacson

18 Girls and Technology 241
Robin Ward

Appendix: Women in Mathematics – Herstory 248

Index 254

Introduction

LEONE BURTON

Girls' participation and achievement in mathematics became an issue in the United States of America in the 1960s but it was not until the 1970s that an equivalent interest began to be demonstrated in the United Kingdom. The ten-year lag allowed those in the UK the opportunity of incorporating the results from the USA into their own thinking while at the same time recognising many differences between the two educational systems. (See, for example, Eddowes and Sturgeon 1981, Walden and Walkerdine 1982.) The passing of the Sex Discrimination Act in 1975 in the UK was a formal recognition of the need for action and attitude change. Now, ten years further on still, is an appropriate time to look at the work which has been done, and to ask in which direction developments will go and what are the implications for schools of our present state of understanding.

Much of the work in America focused on possible explanations for women's poor participation and achievement rates in mathematics. Some of the important contributions which were made were:

(a) Brophy and Good (1974) who did a number of studies on teacher/pupil relationships and their effects on girls' achievement;
(b) Brush (1979) who investigated mathematics avoidance;
(c) Dweck and Bush (1976) who looked at primary school teachers' patterns of interactions with their girl and boy pupils;
(d) Fennema and Sherman (1977, 1978) and Sherman and Fennema, (1977), Sherman (1980) who investigated the effects of attitudes on mathematical performance;
(e) Fox (1979) who was concerned with sex role socialisation;
(f) Horner (1972) who postulated a 'fear of success' as a component in deflecting able girl mathematicians from achieving;
(g) Maccoby and Jacklin (1975) who looked at sex differences in achievement in high school students.

In addition, a number of US investigators subjected to close scrutiny a conjectured link between spatial ability and mathematics. It had been asserted that boys demonstrated greater spatial ability than girls. This assertion led to a hypothesis that the reason for poor female performance in mathematics was genetic rather than environmental.

A recent publication (Chipman, Brush and Wilson, 1985) has brought together this and other work in the United States on participation and achievement. From this work it has been established that the crucial indicators for female participation in mathematics in the secondary and tertiary sector in America are:

(a) perceived utility of mathematics to future career plans;
(b) confidence/enjoyment gained from mathematical experiences;
(c) a supportive social environment within which to learn mathematics.

The major factors affecting these were found to be:

1. Teachers—whose awareness of and sensitivity to the effects of their behaviour on pupils appears to be crucial. Related to this is the kind of classroom environment which is fostered, the image conveyed of the nature of mathematics and mathematical enquiry and the relationships which are developed between the teachers and the taught.
2. Parents—whose expectations of daughters and sons may differ markedly. Girls who continue with an interest in mathematics appear to enjoy more support and encouragement at home especially in the earliest years.

To a large extent, mathematics teachers and families reflect the social context as do other agencies such as the peer group, teachers of other subjects, the media, and so on. But teachers and parents, in the American literature, dominate this scene.

Two persistent assertions about women and mathematics are challenged in this American collection. One is the relationship between spatial ability and mathematical achievement. Susan Chipman and Donna Wilson write:

> There is little evidence either that spatial ability is important to mathematics achievement or that it contributes anything to the explanation of sex differences in mathematics achievement. In general, mathematics enrolment decisions seem to be determined primarily by earlier mathematics achievement, but this is not the explanation of sex differences. Two other influences upon enrolment decisions seem most likely to be the source of sex differences in enrolment. One is the

> perceived utility of advanced mathematics for the student's intended future life and career . . . A second . . . is self-confidence as a learner of mathematics.
>
> (Chipman and Wilson 1985, p. 322)

In a large national study,

> 13-year-old females performed significantly better than males on a test of mathematical computation and performed equally as well as males on tests of problem-solving and algebra as well as overall mathematics achievement . . . females also performed significantly better than males on the test of spatial visualization at age 13.
>
> (Chipman and Thomas 1985, p. 8)

It is during the high school years that any developing sex differences in mathematics achievement are observed in the USA. This is a clear indicator of environmental rather than genetic explanations for such differences.

The second assertion is that women's participation rates in mathematics are poor. Amongst those students continuing to study mathematics at the tertiary level in the United States, approximately 40 per cent are women, and these women are well-prepared with the standard four years of high school mathematics. This figure is drawn from data collected in the latter part of the 1970s and demonstrates an important shift from the findings of Sells (1973), five years earlier, that 57 per cent of Berkeley freshmen in 1972 had taken the standard four years of high school mathematics but only 8 per cent of the women had done so. Susan Chipman and Veronica Thomas focus on

> the problem of girls' participation in high school mathematics courses (which) persists but is not so severe as many people believe . . . in some schools, girls continue to be severely under-represented in mathematics courses. It is now obvious that such a situation is not inevitable and should not be viewed with complacency . . . the problem most evident today (is) the large difference in the number of males and females who obtain very high scores on the . . . maths achievement tests at the end of high school.
>
> (Chipman and Thomas 1985, p. 22)

However, it is salutory to recognise that

> Various groups of minority students are more severely under-represented in advanced high school mathematics courses than female students . . . black, Hispanic and Native American students were only about half as likely to have taken advanced maths courses as white students whereas Asian American students were about twice as likely to have done so
>
> (Chipman and Wilson 1985, p. 323)

and that relatively few students of either sex are majoring in mathematics in the USA.

What relevance does this research have for the British scene? First, we must recognise that the optional nature of mathematics courses in the USA compared to the five compulsory years in British schools represents a difference which is more apparent than real. British schoolgirls continue to be present in the mathematics classrooms but their achievement rates decline. The most recent figures for girls' and boys' participation in public examinations in mathematics indicate that there continues to be differentiated entry to the General Certificate of Education Ordinary level examination at sixteen plus, the girls tending to be entered for the Certificate of Secondary Education instead. Rosie Walden and Valerie Walkerdine point out that, in the opinion of the schools in which they did their research, a CSE Mode 3 which incorporates

> work done in class with that done in examination (gives) a fairer assessment of competence than . . . Mode 1 or 'O' level. However, despite these very valid reasons 'O' level is still the most prestigious examination and more acceptable to future employers. Only Grade 1 CSE is acceptable as an equivalent qualification.
>
> (Walden and Walkerdine 1985, p. 44)
>
> [They] go on to demonstrate that the arguments used by teachers for putting girls in for CSE rather than 'O' level are based upon characteristics which are displayed in the classroom which the teachers tend to interpret as 'lack of confidence'. This has the consequence of making such teachers feel that they should protect rather than push the girls in question . . . we should hardly be surprised to find that, as a consequence, girls' CSE results are not bunched at the top end of the attainment range.
>
> (Walden and Walkerdine 1985, p. 45)

And, indeed, more boys gain the best grades in mathematics in all three public examinations, CSE, 'O' level and 'A' level. In 1983, girls entered for mathematics constituted 4.4 per cent, boys 11.8 per cent of the total 'A' level entry. Of those entered, the results obtained on the first three grades, as a percentage of the entry, were as shown in Table I.1.

Table I.1

	Girls			Boys		
	A	B	C	A	B	C
Pure mathematics	9.79	12.93	12.09	13.57	14.22	9.63
Applied mathematics	12.54	15.46	11.81	16.51	15.21	10.11
Pure and applied mathematics	9.71	13.54	11.48	13.31	13.51	10.65

In the 'O' level examinations, 52.92 per cent of the girls and 61.12 per cent of the boys, gained grades A–C.

In CSE, the entry figures bear out the Walden and Walkerdine contention; 232926 girls were entered, 209973 boys. The pattern of results, however, is interesting (see Table I.2).

Table I.2

CSE grade	1	2	3	4	5	Ungraded
Girls	36033	36458	44591	58431	36474	23939
Boys	38686	34434	41682	47232	29117	18822

The percentage of boys gaining grade 1 (18.42 per cent) is higher than that of girls (15.47 per cent) but the pattern for girls and boys is the same with a skewed distribution peaking at grade 4. These figures are for *all modes* of CSE, with three times as many pupils being entered for Mode 1 as for Mode 3 and only a small number for Mode 2. By focusing on the issue of representation and achievement in examinations, we observe a damaging phenomenon. Our secondary schools are geared to achievement in an examination system at which, demonstrably, our pupils' achievements are not even matching the normal curve of distribution. Clearly, more of the girls are skewed to the bottom end than are the boys. Nonetheless, for many thousands of pupils, five years of mathematics study has left them with a public record of low achievement (statistics from DES 1985).

When the Assessment of Performance Unit (APU) results are examined, interesting consistency is demonstrated with some of the American findings. For example, it is only in the top attainment band that differences in performance become notable. These differences go a considerable way towards explaining the differentiated entry at 'A' level, given that only the most able pupils would be entered for that examination. Lynn Joffe and Derek Foxman, in Chapter 2, state that

> the proportion of boys to girls among 15-year-old pupils obtaining the highest 10 per cent of scores on APU concepts and skills tests is 61 per cent to 39 per cent.

They report on the deteriorating confidence of girls compared to boys in particular demonstrated by attributional patterns (ascribing success to luck, rather than ability) and perceived difficulty (girls having poorer expectations of performance, boys overrating their expected performance). However, one major difference between the

APU and the American findings does have important implications for future work. The APU team is insistent that the differences in performance between girls and boys in the top band are already well-established at eleven years of age and that explanations for this must be sought by looking at the experiences of children before then. Contrary to this, the American researchers claim that it is only in the secondary years of schooling that differences in performance between girls and boys emerge, although Sally Boswell indicates that nine-year-old children 'reveal stereotyped beliefs about mathematics' (Boswell 1985, p.197).

In Chapter 1, Hilary Shuard uses data from the APU primary surveys (APU 1980, 1981, 1982) as well as that from the study *Mathematics and the Ten-year-old* (Ward 1979). In 1983 this highlighted significant differences in performance between girls and boys in the primary school. These differences were explained by looking at what aspects of primary school mathematics considered by the teachers to be important. In her postscript she points out that both the situation and our understanding of it continue to change. She draws attention to the differences in social conditioning of girls and boys which set the parameters within which different teaching styles might influence pupils' experience. For example, will the curious, inquisitive, and independent boy react positively to and learn significantly through an enquiry-based teaching/learning environment which might provoke increased anxiety and discomfiture for the compliant, insecure and unconfident girl? To what degree are these stereotypes of girls and boys borne out by the observations of researchers in the classrooms? Before looking at some evidence, offered in Chapter 10, it is interesting to consider Rosalinde Scott-Hodgetts's contribution to this debate. In Chapter 5, she offers an analysis built upon the Pask distinction between serialistic and holistic learners (1976). Her suggestion is that primary school teachers and primary school content tends to be overwhelmingly serialist and that in conforming to the serialistic role model, female pupils are reinforcing a single strategy method of learning mathematics. She argues that, on the contrary, to learn mathematics requires a flexible approach and a judicious choice of which strategy is appropriate to each mathematical learning task. Such flexibility is only possible, she hypothesises, for those pupils who already use a holistic strategy and for whom the teacher's serialist model presents an alternative. These, she suggests, are more likely to be boys. Her analysis is essentially *post-hoc* in that, by accepting her premises, a credible interpretation of the observations of the APU and of Walden and Walkerdine (1985) can be made. To substantiate this analysis, an empirical study would need to determine the degree to which girls and boys do use these different learning strategies and some hypotheses constructed to identify what aspects of social conditioning steer girls in one direction and boys in the other. The whole analysis has fascinating implications for intervention programmes both in the classroom with pupils and in in-service education courses.

In Chapter 10, Rosie Walden and Valerie Walkerdine assert that there is a continuity through the years of school to age 16 of girls' good performance and positive attitudes. They note that, in general, girls' performance as measured by examination success is at a higher level than boys. The fact that mathematics is not consistent with this general performance points, for all pupils, to factors related to the content of the discipline, the pedagogy through which it is experienced by pupils, and one might add, the social climate and social expectations. But, for girls in particular, they argue that it highlights discontinuities between the way in which their femininity is defined, described and develops and the contradiction thus posed with the discipline and practice of mathematics. One contradiction in particular to which they point is that of the interpretation of differential performance at age 16 which they suggest results from different criteria being used to explain girls' and boys' successes (and failures) in mathematics. A social stereotyping of girls' success ascribes this to hard work and rule-following and this stereotype turns into a self-fulfilling prophecy, in their view. An assumption of lack of confidence or anxiety about mathematics, in turn, both reinforces what is expected and can lead to a *caring* practice by teachers of lessening pressure on girls and offering softer options which, again in turn, ensures poorer performance and increasing anxiety. They assert that

> girls' performance can then be understood as intimately bound up with the criteria for its production and evaluation.
>
> (Walden and Walkerdine 1985, p.17)

The importance of this argument cannot be underrated and future work will be necessary to focus on learning environments which facilitate the construction of knowledge and skills in mathematics together with the powers of reflection and analysis that enable teachers and pupils confidently and competently to assess their feelings and performance. The shift to such a different model of learning would, indeed, represent a major re-assessment of what constitutes knowledge and skills in mathematics and to what degree reproductive patterns of learning are destructive of the very abilities they are theoretically designed to enhance. Within this analysis, it might be said that females could be making a more appropriate and consistent judgement if they reject current curricula and methods in mathematics and that it is the conformity of the males to compete to achieve mastery of codified knowledge which has long-term negative implications. Together with the re-assessment of the content and method of mathematics itself, such a shift also requires a re-assessment of the pedagogical context and climate within which mathematics is taught. This is the issue underlying Chapter 16, which will be explored shortly.

In examining contributions to the debate about participation and achievement in mathematics learning in the UK, we have noted, in particular, the effects on attainment of self-confidence in the learner and of the learning environment. These were both cited as being crucial indicators to attainment in the United States. The third indicator in the USA identified as crucial to attainment is the social environment which carries both complex and variable messages that affect the learning of mathematics. Indeed, Sally Boswell states that

> women's under-representation in mathematics is related to the larger issue of sexism in society.
>
> (Boswell 1985, p.197)

and in Chapter 4, Dale Spender points to the analogy between the power of mathematics as a high-status subject now and the similar role occupied by classical languages in the last century. The same 'problem' of women performing poorly at languages was defined then in the terms now used to describe the 'problem' of poor female performance in mathematics. Meanwhile, the status of language study has altered and so has female performance.

There is much to support in their position. Gilah Leder, in Chapter 6, reviews the historical evidence which suggests that women have, over the centuries, both valued mathematics, mastered it when they had opportunity so to do (despite enormous difficulties sometimes placed in their path and frequently because of support from a significant other) and used their knowledge effectively. However, mathematics retains its image as a masculine domain and she examines some of the societal factors which support this image. She carries out a content analysis on the manner in which the print media portray successful women, in particular those in occupations with mathematical prerequisites. She finds that success for these women is consistently portrayed negatively with emphasis on the need to work harder than men, to jeopardise interpersonal relations and to have a large slice of luck. The hidden message is that a woman enters these domains at her own peril. The assumptions and values which underlie social decision-making are such that power and control rest with males in a masculine context.

In *Women in Mathematics*, Lynne Osen pursues a similar theme:

> We do know that the mysteries and power of mathematics began quite early to come under priestly control, and the mystical characteristics of numbers began to be exploited and emphasized as a valuable religious exegesis. For instance, by plotting the course of the star Sirius, priests were able to predict the annual flooding of the Nile . . . It was no doubt through such usage that mathematics came to be surrounded by an elitism, remnants of which can be found in our culture today.
>
> (Osen 1974, pp.13, 14)

Perhaps this early priestly control was enough to label mathematics as belonging to, and being the product of, a masculine domain. This labelling however has continued to the present and women's attempts to re-define and re-label according to a set of values and assumptions which suit women as well as men are themselves labelled. The norm is defined as masculine so re-definitions must, by that definition, be deviant! For example, in his introduction to *Men of Mathematics* Bell writes that the purpose of the book is 'to see what sort of human beings the men were who created *modern* mathematics' (Bell 1953, p.1) but the reader will find a discussion of Sophie Germain's work in the chapter entitled 'The Prince of Mathematics'. It is prefaced by the remark that the inclusion of the discussion 'shows the liberality of (Gauss's) views regarding women scientific workers' (Bell 1953, p.286). Bell goes on to make reference to, without quoting the name of, Sonya Kovalevsky as 'the most celebrated woman mathematician of the nineteenth century' (p.286). The discussion of her work is found in a chapter entitled 'Master and Pupil'. In a book entitled *Mathematicians and Their Times*, Laurence Young writes of his parents William Henry Young and Grace Chisholm Young, both of whom were mathematicians:

> My parents published, in addition to the *Theory of Sets of Points* (1906), two other mathematical books and 214 papers; 18 papers were my mother's and 13 were joint. This is partly because my mother wanted the credit to go to my father; it would be fairer to say that about a third of the material, and virtually all the writing up of the final versions, was due to my mother . . . To give my father every possible chance, my mother took on the work of a small army of assistants and secretaries; occasionally he slipped in her name as co-author—she certainly did not.
>
> (Young 1981, pp. 282–3)

An extensive discussion of mathematics as a male-defined domain and the effect this had on women's freedom of choice and action is to be found in Chapter 17. In this chapter, Zelda Isaacson draws attention to a number of double-binds facing women in mathematics. One of these is demonstrated in the anecdote quoted above where competency in mathematics, and competency in fulfilling feminine roles, such as wife and/or mother, are seen in a male-dominated society as conflicting. Women must either choose to self-efface as did Grace Chisholm Young, or learn to succeed while withstanding the social pressures of being a woman in a male professional domain and a professional in the domestic female sphere. Changing of this social dimension is dependent upon challenging male attitudes, expectations, fears and insecurities. In this respect, it is just as important that equal opportunities initiatives are taken with males as with females. Those who have looked at attitudes with respect to the sex-role stereotyping of mathematics consistently find

> that although women do indeed hold stereotyped beliefs, males hold them even more strongly.
>
> (Boswell 1985, p.197)

This is often despite an equally clear result in both this country and the United States that primary school and many secondary children claim that girls are just as good at mathematics as boys (see Chapter 2). It is in the secondary years of schooling that failure in confidence, lack of perceived utility of mathematics, poorer attainment in mathematics compared with that in other subjects, an accumulation of social messages and boredom, take their toll.

Social messages about mathematics are being transmitted to young people within the school as well as outside. In Chapter 8, Jean Northam presents her findings on how men and women feature in primary school texts. Given the pervasive presence of such texts, it is a reasonable assumption that the hidden messages which they carry about the roles and functions of women and men are not insignificant for the pupils who are using them. Recently a pack of mathematical games produced in the Inner London Education Authority under the title *Count Me In* drew the comment from teachers examining them: 'Isn't it sad that the fact that in their illustrations they feature all types of children, girls and boys, children from different ethnic groups, handicapped children, hits you because it is so unusual.' Good educational material ought surely to start with that diversity as given.

In Chapter 7, Alison Kelly and her colleagues report on one of the few enquiries that have been conducted in this country into parental aspirations for girls. Contrary to the stereotype, they found that parents were enthusiastic about girls studying physical science and that they had high occupational aspirations and were generally supportive of non-traditional choices for girls.

However frequently these parents were themselves fulfilling sex-stereotyped roles in the family and were making sex-stereotyped demands on their children within the home. The messages being given to these children are, at best, contradictory.

Much more worrying is the work reported upon by Hazel Taylor in Chapter 13. Already in the nursery school, the well-established pattern of permission-seeking from males by females is observable. Equally noted is the phenomenon often described as dependency whereby girls seeking personal interaction with the teacher (for reassurance or security?) even when the activity is a familiar one. Fennema and Peterson (1985) suggest that independent learning behaviour is essential to mathematical development. Dependency is frequently exhibited by females. Thus a neagtive behaviour becomes 'female'. By contrast Gilligan (1982) claims that the criteria for judging dependency/independence are male-derived. However, by far the most disturbing observation was made in the reception class with five-year-old children. The teacher noted different patterns of behaviour and, above all, of task-related talk by girls and boys when building with Lego. The boys regularly chose building as an activity. For the girls, it was a less attractive option if boys were already playing with the Lego. But when they were building, the boys were task-related in their talk, most frequently built moving objects and tended to construct purposefully to fit a global scheme. The girls, on the other hand-built static objects such as houses and used the building time as an opportunity to talk about topics unrelated to the task. So the boys were expanding their spatial experience and their talk was giving them the opportunities to make conjectures and test them in a creative and imaginative way. The satisfaction which they derived, brought them back to the building activity again and again. The girls, meanwhile, were developing their social skills but they were also acting through a pattern which is particularly noticeable in the home. The caring parent discusses with the pre-school child while physically engaged in doing a number of tasks but the tasks rarely form the basis for adult/child talk. More frequently, that talk is carried on despite the task. So by their very versatility in coping with a wide range of tasks and responsibilities simultaneously, caring parents, usually mothers, seem to be presenting a non-task-focused model to their children, a model which the girls are more likely to use than the boys. Hazel Taylor discusses some implications of her observations and describes strategies which were developed for intervening. How much of this learning about behaviour can inform the development of mathematical competencies is a question for the future, but in the construction of images to which children have recourse in later learning, it is foolish to ignore the influences of rich mathematical experience together with the language which enables that experience to be explored and stored. If different patterns of engagement with experience are observed in different sexes, it is only reasonable to assume that what is being made of the experience will, equally, be different.

In this context, it is salutory to examine what is happening to girls and boys in a brand new area of the curriculum but one which has strong messages about links to mathematics and technology. Three chapters discuss the impact of computers and computing on girls, Chapters 9, 11 and 18. They point to a development which has immediately been identified with masculine interests and related to already male-dominated curriculum areas like mathematics, and craft, design and technology, despite the use of (traditionally female) keyboard skills. In the 1983 examinations, two and a half times as many boys as girls entered CSE computer studies, over twice as many boys as girls entered 'O' level and four times as many boys as girls entered 'A' level (DES 1985). Mary Gribbin's and Anita Straker's chapters demonstrate the power of the very early messages that parents are conveying to their children either by what they do and say, or, just as importantly, by what they do not. Such rapid developments alert us to the power of social stereotyping to influence pupils. The computer is a rich and powerful medium for the learning and teaching of mathematics. Negative feelings about computers are bound therefore, to reinforce negative feelings about mathematics, especially since in the present situation of lack of supply of teachers to teach computing and information technology, much of this work is being done by men from mathematics departments in schools, further reinforcing the masculine message.

Reference was made above to the boredom and perceived irrelevance of much of the curriculum offered in mathematics to both girls and boys. The choice of what mathematics constitutes the curriculum has, of course, been made by those in positions of power in the educational world, academics, inspectors, committees of worthies, and, with few exceptions, these have been and continue to be men. Until fairly recently, mathematics was seen in a very utilitarian light by these people, and subsequently did not have the high status it is accorded today. A distinction between mathematics and arithmetic was made in the nineteenth century. Essentially, mathematics was for the middle-class males who had ambitions to enter trade, commerce, the military or engineering. Arithmetic was pursued by girls and all children in the state schools but, for girls, competed with needlework (Cross 1880). But, even recently, books were published written specifically for the teaching of arithmetic and geometry to girls (White and White 1951, Farina 1945). The growth in power and status of mathematics in academic circles was matched to its increasing influence within scientific and technological developments.

> Present-day Western civilisation is distinguished from any other known to history by the extent to which mathematics has influenced contemporary life and thought
>
> (Kline 1953, p.29)

For example, the decision to introduce into school curricula what became known as the 'New' mathematics was a drip down from the positions of influence and power held by pure mathematicians in the universities in the 1950s and early 1960s. The new curriculum emphasised mathematical structures and depended upon developments in pure mathematics, particularly algebra, which had been dominating the university mathematical scene. Nonetheless, despite a description of Emmy Noether as

> the most creative abstract algebraist in the world
>
> (Bell 1953, p.287)

references to her work are not easy to find, especially at school level. In a recent BBC *Horizon* programme called *The Mathematical Mystery Tour*, first shown in December 1984, there was a scroll of famous mathematicians in the history of the subject, all men. In a private communication, the producer of the programme indicated that a reference to Noether had been made in one of the interviews used in the programme but had been edited out! His defence against a criticism of bias in presenting mathematics entirely as a male-dominated subject was that he was reflecting mathematics as it *is* rather than as it *might be*. Thus the reality of women's contribution is re-interpreted to conform to a stereotype which becomes self-perpetuating so the choices on what to include in the curriculum and how to ascribe importance are made by men about men, whether it be at the level of popularisation as described above, or in the writing of texts, or in the preparation of curricula.

In *Sex Differentiation and Schooling*, Michael Marland argues that

> there is virtually no curriculum planning in the United Kingdom

and that

> lack of overall curriculum planning leaves the subjects at the mercy of extraneous influence, and sex stereotyping is one of these
>
> (Marland 1983, pp.144–5)

While not wishing to disagree with his conclusion, it is my observation that the curriculum for mathematics is not planned so much as imposed, usually through the restrictive control of syllabus exercised by a chosen 'scheme' or text series. As Jean Northam shows in Chapter 8 for the primary school, and the work of the Gender and Mathematics Association (GAMMA) texts and resources working group has demonstrated for the secondary school, this imposed curriculum is more often than not sexist. By organising the syllabus in this way, schools block teachers from using the very judgements for which they are trained, judgements relating to teaching styles and strategies, language, learning environment, and so on. Further, there is no possibility to take school-based decisions about the mathematics *curriculum* as most schemes presuppose and prescribe the pedagogical environment in which they will be used. Even if they do not do so overtly, one effect of espousing structure and content through the confident presentation of a commercial scheme is to remove from the teacher the anxiety of having to take these decisions. It is important *not* to diminish the effects of this anxiety in a curriculum area already defined as powerful and masculine. Consequently, teachers willingly abrogate their responsibility to the authors of the schemes. No work has been done as yet on the ways in which curriculum choices reflect masculine interests in society at large, although it is a reasonable assumption that they do. The question of what a mathematics curriculum derived to reflect female as well as male interests might look like is currently being studied and must become more and more important as women's social influence increases. Meanwhile, those concerned to redress the disservice done by the disappearance of women mathematicians from the history of the subject are offered a brief Appendix to this book as a starting point.

The processes through which knowledge is derived impinge upon the content of the curriculum. The strong push in the United States towards introducing problem solving into mathematics courses has recently been endorsed in Britain by the Cockcroft Report recommending both problem solving and investigational work as a necessary part of mathematics teaching (Cockcroft 1982, p.71). Interestingly problem solving in particular has been interpreted in a number of different ways. Much of the American literature deals with using word problems as a basis for problem solving with the emphasis still on the importance of interpretation of the words, abstraction of the required operation, and production of the necessary answer. Apart from the name, it is hard to see in what ways this differs from asking pupils to tackle the set of miscellaneous exercises at the end of a section in a textbook. An alternative interpretation of problem solving, and of investigations as used in the Cockcroft Report, is the construction of a mathematical model to enable pupils to analyse the results of an enquiry into a problematic situation in their environment. Such an enquiry might be conducted on the dinner queue in the school, the traffic outside the school, the running of a sports day, and so on. The emphasis with this approach is on 'best fit' to the situation and the data collected, rather than on correctness, although it quickly becomes clear to the pupils that certain sorts of enquiries and questions are better suited than others.

There is little doubt that the intrusion of any real-world factors into the mathematics classroom is beneficial both to the mathematics and to the pupils but neither the first nor the second approach to problem solving necessarily reflect the kinds of questions or concerns that might be described as female. It is in this context that Chapter 16 is of such great interest. The kind of changed pedagogy to which reference was made above is described by Stephen Brown in this chapter in terms of questioning rather than answering. Posing problems and the opportunity that creates for stimulating discussion and for exploring contradictory and conflicting views on mathematics is a third way to approach the learning of mathematics and one which has many components which feel more comfortable with female styles of working.

First, using language to explore ideas and doing so in a context which encourages alternative interpretations and differing viewpoints is more consistent with the studying of English than Mathematics and consequently more consistent with an area where girls are successful. But the methods are equally appropriate to the growth of understanding of mathematics and offer girls the opportunity to use an area of performance which feels familiar and pleasant.

Second, constructive talking can only happen in small groups which implies that the traditional classroom in which all talk travels between teacher and pupil must change into a classroom in which small groups of pupils engage together on a task and membership of such a group involves a pupil in private rather than public talk. This is well-established as being preferred by women who dislike public confrontation and public ignominy both of which are experienced in the traditional classroom setting. It also means that the dominance of the classroom by boys for social or personal status which has been observed by many researchers no longer becomes an issue.

Third, the mathematics *itself* becomes problematic and open to an investigation of the effects of value choices as detailed by Brown in his chapter. He makes it clear that he perceives a male interpretation of mathematics as involving truth and falsity and the application of logical deduction to a well defined problem area whereas a female approach might be more tentative, focus on differences as well as similarities, ask 'why' as well as 'how' questions, encourage reflection on the implications of making one choice rather than another. Above all, he is making a plea for the mathematics in our classrooms to reflect caring and being responsible (for oneself, for the group, even for the teacher!) rather than, as it more frequently does, engaging pupils in the game of 'what's in teacher's head?' Thus, Brown, Walden and Walkerdine, Taylor and, by implication, many of the other contributors to this book, are moving towards a classroom which builds pupils' independence and autonomy and respects their responsibility to construct their own understanding by challenging their present state of learning. Elsewhere, I have suggested that classrooms could contain a questions board for the collection of questions for which we might or might not have a method of attack (Burton 1984, p.10) but out of which investigations of concern and interest to the pupils can emerge. This is one method of ensuring identification with pupils' current levels of awareness and of counteracting boredom and disaffection in girls and boys. Little has so far been done to investigate what effects such changes in the mathematics classroom might have on girls' and boys' attitudes to mathematics and attainment in the subject. As more and more teachers adopt mathematical problem solving in some form as a part of their teaching strategies, the need to establish what differences the three approaches outlined above make to performance will become more pressing. Such work as has been done indicates that shifts in attitude do take place once it is understood that the pupils are trusted to make their own decisions and conduct their own enquiries and the time wasted by teachers dealing with disruptive boys and alienated girls can be redirected towards more professionally rewarding behaviour. Barbara Binns reinforces this message from a classroom perspective in Chapter 12.

It seems that the 'problem' is not one of the lack of girls' participation in mathematics so much as the nature of mathematical experiences and curriculum offered to all pupils and the way in which mathematics is defined in relation to the sexist distribution of power and authority in society. Intervention programmes which attempt to remediate the 'girls' problem' without understanding that it reflects a societal situation are unlikely to be effective. Consequently, looking at gender issues in the teaching of mathematics becomes one way of looking at the teaching of mathematics and asking in what ways it succeeds and fails with all pupils. Many of those writing in this collection point to aspects of the teaching and learning process which, if adjusted to coincide with the needs and interests of the female pupils, are more likely to represent better practice for all. Other insights could be obtained by subjecting the teaching and learning of mathematics to scrutiny through lenses designed to highlight for example, multiculturalism or issues about handicaps. In each case, we can only expand and inform our understanding of the difficult and complex processes through which mathematics is learnt. I feel sure, for example, that a knowledge of the particularities of teaching mathematics to the deaf would provide a highly informative perspective on the teaching of all pupils. The literature of mathematics education has failed until now to take such particular insights and place them in a general pedagogic context. This volume attempts to do so from a gender perspective. The intention, therefore, is to carry meaning not only about ways in which discriminatory practices affect the mathematics learning of girls in particular, but to invite the reader to continue to reflect on the implications of these for her, or his, teaching and learning strategies in general.

Chapters 1 to 10 constitute the first section of the book, What is the problem? They raise issues relating to social practices such as sex-role stereotyping, social expectations reflected in parental or teacher behaviour, social attitudes towards, for example, mathematics as a male domain and the impact these have on attainment in mathematics. The uneasy relationship between mathematics and computing has not been ignored. Chapters 11 to 18 constitute the second section, What can be done? All these contributions address themselves to the need to redefine pupils' experience of mathematics and computing in some way, whether by adopting a whole school approach to equal opportunities as described in Chapter 14 by Alan Eales, or by running mathematics and career awareness days such as the Be A Sumbody conferences described in Chapter 15. That action is necessary throughout the school system from nursery to tertiary is made clear by the spectrum of contributions from all sectors. The Appendix, 'Women in Mathematics—Herstory', provides a first resource to those who would like to introduce pupils to some women mathematicians who have made, or continue to make, contributions to the development of the subject. One way of broadening the scope of the discipline, and developing pupils' relations to it, is to follow up this beginning, and encourage pupils to research information about women mathematicians and the mathematics they developed. Continuing such work into the present day, and inviting women in mathematical, technical and scientific jobs to meet pupils, forges links between mathematics learning, social utility and social expectations and changes the expectations of both girls and boys.

The book was conceived as support to an in-service pack (PM645) entitled 'Girls into Maths' which has been jointly developed by the Inner London Education Authority and the Open University and is available from the Open University. The contributions were deliberately chosen from a wide range of sources, some by academics, some by practitioners, some written for a general readership. Together they present a current picture of the state of thinking about gender issues in mathematics education, without claiming to be all-inclusive. It is hoped that every reader can find an entry point into these issues through one of the contributions and that this will encourage further exploration.

The distinction between the two sections 'What is the problem?' and 'What can be done?' is by no means as clear as the separation might suggest. Interest in identifying the impact and the results of sex-differentiation in mathematics teaching and learning automatically implies interest in affecting current practice. Overall, the contributors share a commitment to enhancing pupils' learning potential. Much of what we are learning from a feminist perspective on learning and teaching mathematics informs general issues. For example, the issue of assessment is affected by analyses done on sexually differentiated responses to multiple choice and written examinations; the issue of learning style is affected by perspectives on group learning which is collaborative rather than individual or class-based learning which can be competitive.

The importance of *now* is that there is a critical mass of work being done by those practitioners and researchers with a feminist interest. The learning potential of all pupils will be affected. It is no longer necessary to ask 'Is there anybody out there listening?' because 'out there' is now in the schools, classrooms and experiences of our girls and boys.

REFERENCES

Assessment of Performance Unit (APU) (1980) *Mathematical Development*, Primary Survey Report No. 1. London: HMSO.

Assessment of Performance Unit (APU) (1981) *Mathematical Development*, Primary Survey Report No. 2. London: HMSO.

Assessment of Performance Unit (APU) (1982) *Mathematical Development*, Primary Survey Report No. 3. London: HMSO.

Assessment of Performance Unit (APU) (1985) *A Review of Monitoring in Mathematics: 1978 to 1982*. London: HMSO.

Bell, E.T. (1953) *Men of Mathematics*. Harmondsworth: Penguin.

Boswell, S.L. (1985) 'The influence of sex-role stereotyping on women's attitudes and achievement in mathematics' in Chipman, S.F., Brush, L.R. and Wilson, D.M. (eds) *Women and Mathematics: Balancing the Equation*. Hillsdale, NJ and London: Lawrence Erlbaum Assoc.

Brophy, J.E. and Good, T. (1974) *Teacher–Student Relationships: Causes and Consequences*. New York: Holt, Rinehart and Winston.

Brush, L.R. (1979) *Why Women Avoid the Study of Mathematics: a Longitudinal Study*. Cambridge, MA: ABGT Associates, 1979 (ERIC Document Reproduction Service No. ED 188 887).

Burton, L. (1984) *Thinking Things Through*. Oxford: Basil Blackwell.

Chipman, S.F., Brush, L.R. and Wilson, D.M. (1985) *Women and Mathematics: Balancing the Equation*. Hillsdale, NJ and London: Lawrence Erlbaum Assoc.

Chipman, S.F. and Thomas, V.G. (1985) 'Women's Participation in Mathematics: Outlining the Problem' in Chipman, S.F., Brush, L.R. and Wilson, D.M. (eds) *Women and Mathematics: Balancing the Equation*. Hillsdale, NJ and London: Lawrence Erlbaum Assoc.

Chipman, S.F. and Wilson, D.M. (1985) 'Understanding Mathematics Course Enrolment and Mathematics Achievement: A Synthesis of the Research' in Chipman, S.F., Brush, L.R. and Wilson, D.M. (eds) *Women and Mathematics: Balancing the Equation*. Hillsdale, NJ and London: Lawrence Erlbaum Assoc.

Cockcroft, W.H. (Chair) (1982) *Mathematics Counts*, Report of the Committee of Inquiry into the Teaching of Mathematics in Schools. London: HMSO.

Cross (Chair) (1880) (The Cross Report) Final Report of the Commissioners Appointed to Inquire into the Elementary Education Acts, England and Wales. London. HMSO.

DES (1985) Statistics of Education, School Leavers CSE and GCE, England 1983. London. HMSO.

Dweck, C.S. and Bush, E.S. (1976) 'Sex differences in learned helplessness', *Developmental Psychology*, **12**, 147–561.

Eddowes, M. and Sturgeon, S.B. (1981) *Mathematics Education and Girls: a Research Report*. Sheffield City Polytechnic and the British Petroleum Company Ltd.

Farina, E.J. (1945) *Geometry for Women's Trades*. London: Pitman.

Fennema, E. and Peterson, P. (1985) 'Autonomous Learning Behaviour: A Possible Explanation of Gender-Related Differences in Mathematics'. In Wilkinson, L.C. and Marrett, C.B. (eds) *Gender Influences in Classroom Interaction*, Orlando: Academic Press.

Fennema, E. and Sherman, J. (1977) 'Sexual stereotyping and mathematics learning', *The Arithmetic Teacher*, **24**(5), 369–72.

Fennema, E. and Sherman, J. (1978) 'Sex-related differences in mathematics achievement and related factors: A further study'. *Journal for Research in Mathematical Education*, **9**, 189–203.

Fox, L.H. (1979) *Women and Mathematics: the Impact of Early Intervention Programs upon Course Taking and Attitudes in High School.* Baltimore: John Hopkins University (ERIC Document Reproduction Service No. ED 188 886).

Gilligan, C. (1982) *In a Different Voice.* Cambridge, Mass. and London: Harvard University Press.

Horner, M. (1972) 'The motive to avoid success and changing aspirations of college women' in Bardwick, J. (ed.), *Readings in the Psychology of Women*, New York: Harper and Row.

Kline, M. (1953) *Mathematics in Western Culture.* Oxford: Oxford University Press.

Maccoby, E.E. and Jacklin, C.N. (1975) *The Psychology of Sex Differences.* Oxford: Oxford University Press.

Marland, M. (ed.) (1983), *Sex Differentiation and Schooling.* London: Heinemann Educational.

Osen, L. (1974) *Women in Mathematics.* Cambridge, Mass. and London: MIT Press.

Pask, G. (1976) *The Cybernetics of Human Learning and Performance.* London: Hutchinson.

Sells, L. (1973) 'High school mathematics as the critical filter in the job market. Developing opportunities for minorities in graduate education.' 47–39. Proceedings of the Conference on Minority Graduate Education at the University of California, Berkeley.

Sherman, J. (1980) 'Mathematics, spatial visualization and related factors: Changes in girls and boys, grades 8–11.' *Journal of Educational Psychology*, **72**(4), 476–82.

Sherman, J. and Fennema, E. (1977) 'The study of mathematics by high school girls and boys: Related variables.' *American Educational Research Journal*, **14**, 159–68.

Walden, R. and Walkerdine, V., (1982), *Girls and Mathematics: The Early Years.* London: University of London Institute of Education Bedford Way Papers 8.

Walden, R. and Walkerdine, V. (1985) *Girls and Mathematics: From Primary to Secondary Schooling.* Bedford Way Papers 24. London: University of London Institute of Education.

Ward, M. (1979) *Mathematics and the Ten-year-old.* London: Evans/Methuen Educational.

White, W.B. and White, E. (1951) *Essential Everyday Arithmetic for Girls.* London: University of London Press.

Young, L. (1981) *Mathematicians and Their Times.* Amsterdam and Oxford: North Holland Publishing Company.

PART I

What is the Problem?

1

*The Relative Attainment of Girls and Boys in Mathematics in the Primary Years**

HILARY SHUARD

INTRODUCTION

It is generally thought that there is little difference between the mathematical attainment of girls and boys at the primary stage; the greater success which boys display later, when they take public examinations in mathematics, is thought to appear at about puberty, and then to build up as the pupils grow older. This view was well expressed by Lesley Kant, of the Schools Council, when I wrote to her in 1980 to make enquiries on behalf of the Cockcroft Committee:

> Although sex differences in relation to mathematical performance are minimal at primary level, I am convinced that socialisation and conditioning can influence children's attitudes to subjects and consequently affect learning patterns and performance at a later stage.
> (Kant, 1980, private communication)

In this paper, I shall draw attention to some evidence which supports the view about pupils' *attitudes* to mathematics which Lesley Kant expressed. Other evidence will also be discussed; this suggests that, where children's *performance* in mathematics is concerned, the situation is very complicated during the primary years. It seems likely that boys' and girls' relative mathematical performance depends on the types of questions which they are asked, even at this early stage in their education. Hence, evidence about boys' and girls' relative

*This chapter is an expanded form of a paper presented to a conference of the Girls (now Gender) and Mathematics Association (GAMMA) in May 1983 and published in the Gemma Newsletter, **4**, Sept. 1983.

performance may depend on the balance of topics found in their curriculum, and the tests used to examine their attainment. I believe that Muriel Eddowes' summary of the effects of early experience, in her book *Humble Pi*, is still oversimplified:

> A major difference between the sexes in primary school is boys' superiority in spatial and mechanical tasks. Since this appears to give a firmer base for mathematics and science at secondary school than girls' superiority in computation, the interaction of young children and their toys could be a crucial factor for later mathematical progress. The effects on mathematical performance of these differential experiences do not tend to appear until secondary level. The girls' superiority at this age in verbal and basic computational skills carries them through the assessments at primary level.
>
> (Eddowes 1983, 16–17)

I shall suggest that by the third year of the junior school, boys are already ahead at those aspects of mathematics which are fruitful for future mathematical development and insight. If a mathematics test favours computation rather than problem solving, it may give an inflated view of a girl's all round mathematical attainment, and of her potential for future success.

ATTITUDES

For the past five years, the Assessment of Performance Unit (APU) has been monitoring the performance in mathematics of large samples of 11-year-old and 16-year-old pupils in England and Wales. Little difference was found between girls' and boys' enjoyment of mathematics, or their perception of its usefulness, but there were significant differences in the opinions of girls and boys on how *difficult* they found mathematics. Girls responded with significantly more agreement than boys to these statements:

> I often get into difficulties with my mathematics. I'm surprised if I get a lot of maths right. Maths often gets too complicated for me.

Boys, on the other hand, agreed significantly more than girls with the following statements:

> I usually understand a new idea in mathematics quickly. Maths is one of my better subjects.
> I usually get most of my maths right.
> I don't think maths is difficult.

Thus, it would seem likely that when children enter their secondary schools, more boys than girls arrive in their new schools expecting to be successful in mathematics, while many girls already feel that they are failing at the subject.

Feelings seem to be independent of any hard evidence from the pupils' actual mathematical performance. The percentage of 11-year-old girls who agreed that they 'often got into difficulties with their maths' was 54 per cent, and 59 per cent were surprised if they 'get a lot of maths right'. The corresponding figures for boys were 46 per cent and 51 per cent.

The APU summed up its findings as follows:

> As in previous years, girls concurred more strongly than boys with statements expressing opinions about the difficulty of mathematics. Boys expressed more confidence in their own ability and greater expectation of success. They also expressed greater confidence in relation to various mathematical topics and activities, particularly those involving measurement. Girls were less certain of the degree of difficulty presented by such activities.
>
> (APU 1982, p.97)

ATTAINMENT AT THE AGE OF TEN

Some very useful material for studying the relative attainment of girls and boys during the primary years is contained in a Schools Council project undertaken by Murray Ward in 1972–75; the project was 'Primary School Mathematics: Evaluation Studies', and its report was entitled *Mathematics and the Ten-year-old* (Ward 1979). This study focused on children in the third year of junior schools in England and Wales. Amongst other activities undertaken by the project, a national sample of about 2300 children were given pencil-and-paper tests in mathematics (the test was given in four versions, so that every question was answered by 500 or more children); girls' and boys' successes on each test item were reported separately. The project team did not themselves specifically study the relative performance of girls and boys, but they published their results in a way which makes it possible to use them for this purpose. I am most grateful to Murray Ward and his colleagues for providing such a wealth of detailed information. The test questions reflected a broad mathematical curriculum:

> Eventually a framework was produced giving broad topic areas—computation, number structure and so on—each being broken down into a hierarchy of concepts . . . We developed from the framework a list of mathematical topics that 10-year-old children might be

> concerned with, such topics stretching as widely across the subject as possible to encompass 'traditional' and a variety of 'progressive' backgrounds. We did not expect that any one class would have covered all the topics, but that every child should have encountered a majority of them . . . In some (questions) the child is told the mathematical operation required (for example, by being told to add), so here we are testing his computational skill. In others—and we usually call these 'problems'—the child has to read the question and reason out what mathematics is involved.
>
> (Ward 1979, p.33–5)

In over a quarter of the test questions, there was a significant difference between the performance of girls and boys; in 11 of the 91 questions, girls did significantly better, while in 14 other questions, boys did significantly better. The items in which these significant differences occurred were not scattered over the whole curriculum, but they clustered into certain topics.

Six of the questions at which girls were significantly more successful were concerned with *computational skills* with whole numbers and money. Figure 1.1 illustrates these questions, together with the percentages of girls and boys who carried them out correctly, and the probability that such an effect would occur by chance. None of the questions at which boys were significantly more successful was a straight computational skill question. The only question which looks at first sight to be a test of computational skills is the question shown in Figure 1.2. In fact, this question tests pupils' understanding of the relation between multiplication and division, rather than testing their ability to carry out a computation. Girls were also significantly more successful at three questions which were set in purely verbal terms, together with a question in which a pattern had to be followed, and one on the relationship between two sums of money. These questions are shown in Figure 1.3. Much other research has found that girls tend to score higher on tests of *verbal ability* (Maccoby and Jacklin 1975).

Four of the questions on which boys performed significantly better were concerned with number; they were classified by the project as testing children's *understanding of place value* concepts. They are shown in Figure 1.4. Boys also did better than girls, but not significantly so, on the other items which tested understanding of place value. The remaining items at which boys did significantly better are shown in Figure 1.5; they are concerned with *measurement*, *spatial visualisation* and *problem solving*. Again, it is well known that boys perform better than girls at tests which involve spatial visualisation (Maccoby and Jacklin 1975).

Question 1

Add 349 + 264 = ____

Girls' success rate = 92%
Boys' success rate = 86%

($p < 0.01$)

Question 3

Add £3.30 + 0.35 + 0.75 = ____

Girls' success rate = 81%
Boys' success rate = 73%

($p < 0.01$)

Question 10

315 ÷ 6

Write down and work out this division sum.

Girls' success rate = 56%
Boys' success rate = 44%

($p < 0.01$)

Question 26

Write the number missing in this subtraction sum:

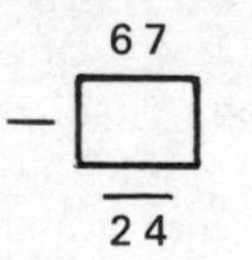

67 − ☐ = 24

Girls' success rate = 80%
Boys' success rate = 73%

($p < 0.01$)

Question 8

Multiply 283 × 7 = ____

Girls' success rate = 58%
Boys' success rate = 51%

($p < 0.05$)

Question 28

Write the number missing in this addition sum:

43 + ☐ + 24 = 99

Girls' success rate = 75%
Boys' success rate = 69%

($p < 0.05$)

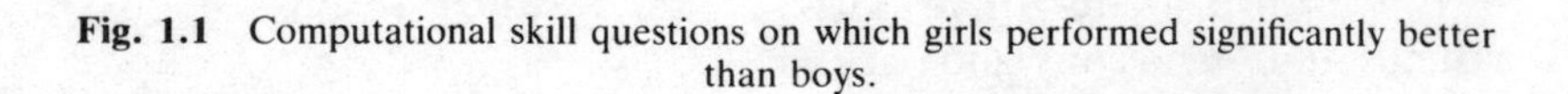

Fig. 1.1 Computational skill questions on which girls performed significantly better than boys.

Question 33

$$105 \div \nabla = 21$$

What number does ▽ stand for?

Girls' success rate = 37%
Boys' success rate = 46% ($p < 0.05$)

Fig. 1.2 A 'computation' question on which boys performed significantly better than girls.

Questions 64–8
Under each shape write its name. The first has been done for you.

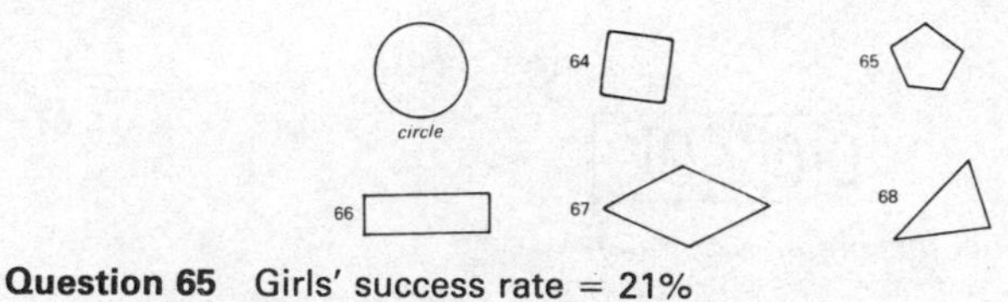

Question 65 Girls' success rate = 21%
Boys' success rate = 14% ($p < 0.05$)

Question 67 Girls' success rate = 78%
Boys' success rate = 69% ($p < 0.05$)

Question 91
Simon is shorter than Trevor, and Trevor is taller than Vic. Put a tick in the box under Trevor.

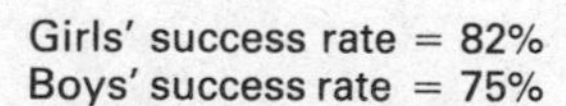

Girls' success rate = 82%
Boys' success rate = 75% ($p < 0.05$)

Question 79
Points on the top line are joined to points on the bottom. Draw two more arrows showing the same relation:

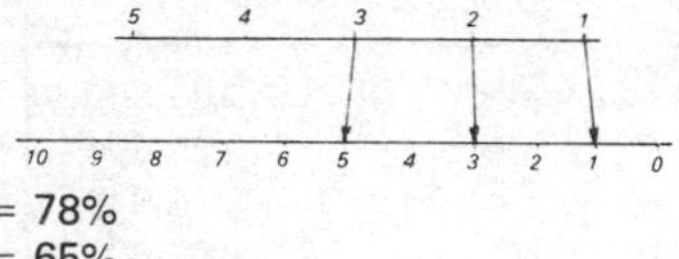

Girls' success rate = 78%
Boys' success rate = 65% ($p < 0.001$)

Questions 34–6
Write $<$, $>$, or $=$ in the circles to make these correct:

34 3 × 2 ○ 3 + 4

35 90p ○ £1

Question 35 Girls' success rate = 43%
Boys' success rate = 34% ($p < 0.05$)

Fig. 1.3 Other questions on which girls performed significantly better than boys.

Question 43
The table shows approximate populations:

Aberdeen	183 800
Bath	151 500
Fleetwood	28 800
Walsall	184 600
Winchester	31 000

Which one of these towns has the largest population?

Girls' success rate = 58%
Boys' success rate = 72% ($p < 0.001$)

Question 13

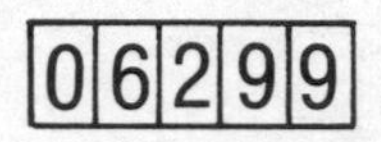

What is the next number this meter will show?

Write it in here 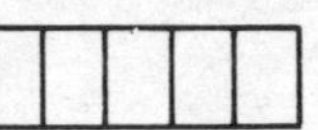

Girls' success rate = 34%
Boys' success rate = 47% ($p < 0.001$)

Question 14
The milometer on a car shows 0 2 6 9 9 *miles.*

What will it show after the car has gone one more mile?

Girls' success rate = 42%
Boys' success rate = 54% ($p < 0.001$)

Question 42 *The table shows approximate populations:*

Aberdeen	183 800
Fleetwood	28 800
Hitchin	26 900
Walsall	184 600
Winchester	31 000

Which one of these towns has the smallest population?

Girls' success rate = 58%
Boys' success rate = 72% ($p < 0.05$)

Figure 1.4 Questions testing understanding of place value at which boys did significantly better than girls.

Questions 88–90
Underline the best guess for each of these:

88 *How tall is a pint milk bottle?* 2 cm 20 cm 200 cm 2 m
89 *How heavy is one drawing pin?* 1 g 10 g 100 g 1 kg
90 *How high is a table?* 1 cm 10 cm 1 m 10 m

Question 90 Girls' success rate = 48%
Boys' success rate = 64% ($p < 0.001$)

Questions 50–52 *If these clocks are all ten minutes fast, write the correct time under each one:*

50

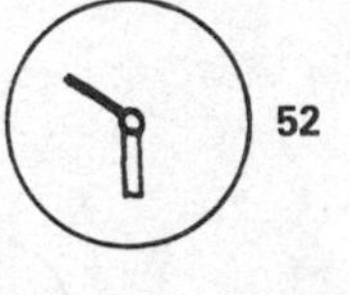

Question 51 Girls' success rate = 43%
Boys' success rate = 57% ($p < 0.001$)

Question 61 *This is the plan of a room. What is the area of its floor (in square metres)?*

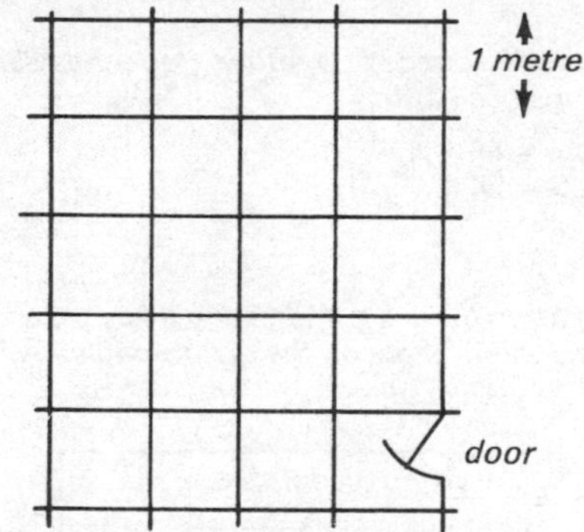

Girls' success rate = 28%
Boys' success rate = 41% ($p < 0.001$)

Questions 59 and 60 *This chart shows the height of five children:*

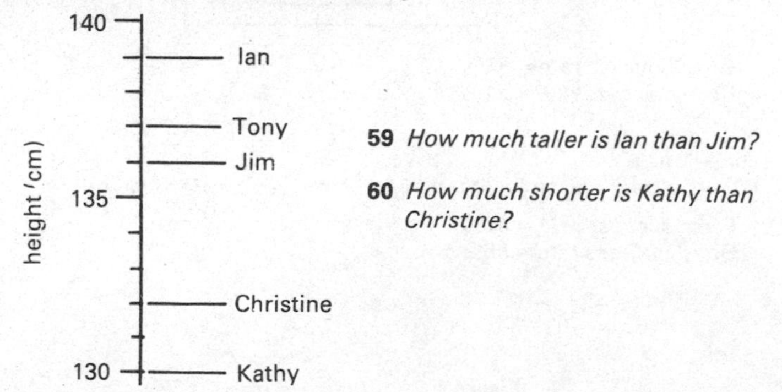

59 *How much taller is Ian than Jim?*

60 *How much shorter is Kathy than Christine?*

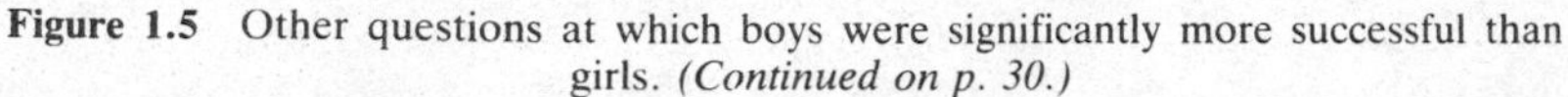
Figure 1.5 Other questions at which boys were significantly more successful than girls. *(Continued on p. 30.)*

Question 59 Girls' success rate = 40%
Boys' success rate = 49% ($p < 0.05$)

Question 60 Girls' success rate = 46%
Boys' success rate = 58% ($p < 0.01$)

Question 71
If you are facing north and turn clockwise through a right angle, what direction will you be facing then?

N

W *E*

S

Girls' success rate = 59%
Boys' success rate = 70% ($p < 0.01$)

Question 48
Paul and Roger have 40 stamps altogether. Roger has 3 times as many as Paul, so how many stamps has Paul?

Girls' success rate = 44%
Boys' success rate = 54% ($p < 0.05$)

Question 81
In a group of 10 children there are 4 girls and 6 boys. 3 of the girls and 4 of the boys wear glasses. Put numbers on the diagram to show where all these children fit in.

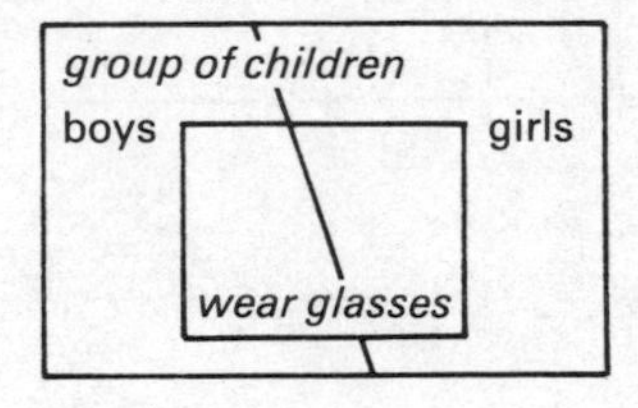

Girls' success rate = 14%
Boys' success rate = 21% ($p < 0.05$)

Question 56
60% of a class can swim. What percentage are unable to swim?

Girls' success rate = 47%
Boys' success rate = 56% ($p < 0.05$)

Figure 1.5 *(Continued from p. 29.)*

In a questionnaire sent to teachers taking part in the Schools Council project, teachers were asked to rank many of the test questions given to pupils in their order of importance:

> How important do you think it is that a 10-year-old child of average ability should be able to answer questions like the following? Each is an example of a *type* of problem.

It was thought that this question would provide some indication of the emphasis which teachers placed on particular topics when they were teaching mathematics—a topic which a teacher thought very important would naturally receive a good deal of emphasis in the teacher's classroom work. It was found that the questions at which girls did signficantly better were ranked by teachers as being more important than those at which boys did significantly better. The mean ranking of the 'girls' questions' was 10.9, while that of the 'boys' questions' was 16.4 (the most important question of all would be ranked 1 by a teacher). Thus, it seems likely that the *questions at which girls did better were more emphasised in the work which the children do in class*. Moreover, the *questions at which girls did signficiantly better were easier questions* than those at which boys did significantly better; the mean success rate of the 'girls' questions' was 64 per cent (for boys and girls together), while the mean success rate of the 'boys' questions' was only 49 per cent (for boys and girls together). The hypothesis that the primary mathematics curriculum favours boys does not seem to be well supported by this evidence. The four questions which teachers thought to be of the greatest importance are shown in Figure 1.6; in questions on three of these four types, girls did significantly better. It would be mischievous to suggest that pupils who pay attention to the teacher's traditional emphases in primary mathematics give themselves a positive *dis*advantage for future success in mathematics, but the evidence seems to point in this direction.

348	349	392	£3.30
−193	+264	× 7	0.35
			+0.75

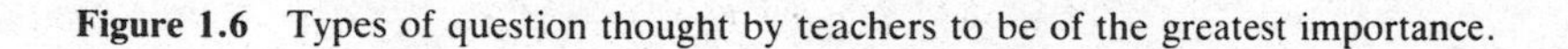

Figure 1.6 Types of question thought by teachers to be of the greatest importance.

ATTAINMENT AT THE AGE OF ELEVEN

In the report of the third APU primary survey (APU 1982) there is a discussion of the pattern of results found in the first three surveys. It is noted that the rank order of the differences between boys' and girls' scores on different mathematical topics has remained substantially the same over the three surveys. The combined order is listed below; those topics at which boys excel most are listed at the top of the list, and those at which girls have the greatest advantage over boys at the bottom of the list. Only in the last two topics did girls perform better than boys in all three surveys.

(a) applications of number;
(b) length, area, volume, capacity;
(c) rate and ratio;
(d) concepts of fractions and decimals;
(e) computation with fractions;
(f) money, time, mass, temperature;
(g) concepts: whole numbers;
(h) symmetry, transformation, co-ordinates;
(i) probability and data representation;
(j) lines, angles, shapes;
(k) sets and relations;
(l) generalised arithmetic;
(m) computation: whole numbers and decimals.

In each survey, up to five of the differences were significant. It is noteworthy that girls were more substantially ahead on 'computation: whole numbers and decimals' than on any other topic. The topic of 'generalised arithmetic', on which girls also did better than boys, seems to be fairly closely related to computation, and to a search for patterns. We might speculate that computational algorithms are based on patterns. Two examples from the topic of 'generalised arithmetic' are given below; separate results for girls and boys do not appear in the first and second reports.

Find which number Δ stands for.

$51 + \Delta = 90$ $\Delta =$ ....

Fill in the values of M in this table according to the equation $M + N = 4$

N	0	1	2	3	4
M	4				

Figure 1.7 'Generalised arithmetic' examples.

Little detail has yet been published about the topic 'applications of number', at which the difference between girls and boys is most in favour of boys. However, the following are two examples of questions in this category from the first survey:

> Spoons are sold in boxes containing half a dozen.
> If I want 30 spoons, how many boxes shall I buy?
>
> 256 children are going to have tea at the Christmas party. 8 children can sit at a table.
> How many tables will be needed?
>
> (APU 1980)

These items would usually be classified as 'word problems'. Other types of questions at which boys did relatively well on the APU's tests seem to be those involving measurement and spatial visualisation, as has often been found in other research.

An interesting example of the differences between girls and boys in a practical situation is given in the second primary survey (APU 1981). In the individual practical tests, children were asked to carry out tasks involving weighing. They were provided with a balance (but no weights), and were asked to use the balance to determine which of three given objects was the heaviest. Most 11-year-old boys (84 per cent), but only 69 per cent of girls, were able to do this. Later, the children were given a 20-gram weight and a bag of identical small plastic shapes; they were asked to use the balance to find how much one of the shapes weighed. This task was correctly carried out by 57 per cent of boys, and 40 per cent of girls. The report comments: 'Almost twice as many girls as boys initially attempted to balance a single plastic shape against the 20-gram mass.' In a stereotyped analysis, it might have been thought that weighing was a practical task which more girls than boys might have experienced, through helping and watching their mothers at work in the kitchen.

More detail than can be provided at this global level of 'topic areas' is certainly needed, if much of significance is to emerge. For example, the topic of 'money, time, mass, temperature' contains 'money', where it looks possible that girls may do better, and 'mass', where there seems to be some evidence that boys find the topic more manageable.

CONCLUSION

Much still remains to be done in understanding the detail of girls' and boys' relative performance at the primary stage. Until differences are established at a detailed level, it seems to me to be very unlikely that efforts intended to prevent girls from being left behind in mathematics will succeed.

As children progress through the school curriculum, some mathematical topics become more and more important, and others become less and less important. Computation is a prime example of a topic whose relative importance declines sharply as pupils get older, and especially as their mathematics becomes more advanced. Technology is also changing the importance of computational skills; the skills of pencil-and-paper calculation are becoming much less important as calculators and computers do more of our computation, in place of the human clerks who used to have to perform this hack work a few decades ago.

As pupils grow older and tackle more advanced mathematics, problem solving and the understanding of mathematical concepts become more important, so that they are absolutely central to progress. Somehow, by the age of ten, more boys than girls have got themselves into a position where they are able to cope with these aspects of mathematics. Judging by Ward's work (1979), it does not seem very likely that the experiences which boys have in their primary school mathematics lessons are responsible for this fortunate bent, because the majority of primary teachers do not seem to regard the problem solving aspect of mathematics as important, compared with the skills of computation, at which more girls do well.

Two pitfalls seem to me to open up when we attempt to correct the imbalance. First, a generalised attempt to get girls to try harder at mathematics in the primary school may produce exactly the opposite of the intended effect, because girls may put more effort into excelling at computational skills rather than at understanding the concepts and principles of mathematics. Secondly, in those classrooms where, as the Cockcroft Report (1982) recommends, mathematics involves not only the practice of skills, but also practical work and problem solving, discussion and investigation, there is the danger that the style of the mathematics lessons will suit the social conditioning of the boys rather than that of the girls, and boys will benefit from the more varied mathematical diet more than will the girls. Certainly, if children are to prepare for future success in mathematics, it is vitally important that their mathematics should include practical work, problem solving, discussion and investigation. However, it is equally important that girls should be urged to take a full part in those activities, if they are to improve their mathematical understanding. Only by building a basis of understanding can girls expect to be able to continue to keep up with the mathematical work of the secondary school, and to use their mathematics as they need to, in their higher education and in their future careers.

POSTSCRIPT

The above paper was written and delivered to the GAMMA conference in 1983. Since then, some new work has appeared which relates to the relative attainment of girls and boys in mathematics in the primary years. The most important is the detailed retrospective review of the APU's monitoring of mathematics over the five-year period 1978–82 (APU, 1985). This substantial, two-volume report contains further evidence related to gender. In summary, the APU confirms that differences continue to be found in the performance of girls and boys at the age of 11. However, these differences vary substantially between different topic areas, and are consistently significant in each of the five years in some topics.

> Boys are best relative to girls in the applied and practical area – Measures and Rate and ratio – while girls do best relative to boys in Computation (whole numbers and decimals) and Algebra . . . At age 11 the only comparisons between boys and girls within a given subcategory to show significant differences consistently are in Applications of number and Rate and ratio.
>
> (APU 1985, p.820)

However, further analysis has revealed that these differences do not occur equally in all the attainment bands; it is found that nearly all the differences are accounted for by the much better performance of the boys at the upper end of the range of attainment – in 1981 and 1982 the top ten per cent of attainers at age 11 was made up of 58 per cent boys and 42 per cent girls. In other attainment bands the performance of boys and girls was much more similar. Thus, the APU concludes:

> For the large majority of boys and girls, there is a great deal which is similar in their mathematics performance.

However, the difference in the top band gives cause for much concern, as a pupil who is in the top 20 per cent at the age of 11 does as well on the APU's tests as do pupils of average attainment or higher and five years older, when they do the same items. Thus, those who are performing very well in mathematics at the age of 11 must have a very good chance of continuing to do so during the secondary years.

There are also some indications that the differences between the performance of girls and boys are widening rather than narrowing as time goes by. In its study of trends in performance over the five-year period 1978–82, the APU has identified significant changes in the relative performance of boys and girls at the age of 11 in two subcategories of content; these are the 'measures' subcategory of length, area, volume and capacity, and the subcategory of rate and ratio. In the measures:

> Boys are performing on average about 2½ per cent better in 1982 than in 1978 whereas the girls are only about 1 per cent better. The average difference between boys and girls goes up from 2.8 per cent in 1978 to 4 per cent in 1982; boys are improving at a faster rate than girls.

A similar result was obtained in rate and ratio; the boys were performing about 6 per cent better than the girls in 1982, whereas they were about 4 per cent better in 1978.

By 1984, Murray Ward's data was ten years old; although his results were published in 1979, he carried out his tests in 1974, and since it seems that changes are taking place gradually over time, it might be questioned whether Murray Ward's data is now of any value. However, in 1984 I was able to replicate Murray Ward's test, in one of the four versions which he used, Version P, with 477 third-year junior children, made up of 240 boys and 237 girls, in schools around Cambridge in which a member of staff was attending an INSET course I was organising. The size of the sample was comparable with that used by Murray Ward – 579 children did Version P of the test in 1974 – but the sample of schools used in 1984 was not random, as it was in 1974. However, only a minority of the teachers attending the INSET course were actually teaching the children tested.

The 1984 test showed some changes in overall performance (Shuard, in preparation), consistent with a slow change in the emphasis of the curriculum from formal computational algorithms to a greater prominence for graphical and visual representation. The test also included questions in the measures which involved diagrams. There was a decline in overall success rate for boys and girls together of about three per cent in performance on straightforward computational questions, but an increase ranging from two per cent to 16 per cent on questions involving graphical and diagrammatic representation, including questions in area and perimeter.

It might be expected that such a change in curriculum might itself disadvantage girls, since many of their past successes have been on the computational algorithms. The actual results show more complexity in their make-up than this, but the girls have made substantial gains (more than three per cent) relative to boys over the ten-year period on five items of the 31-item test, while boys have made similar substantial gains relative to girls on ten items. However, two of the five items on which girls made substantial gains were concerned with the reading of analogue clock faces; there was in fact a considerable falling-off in the performance of both sexes, but this was much greater in the case of boys. Another enquiry of the same 477 children revealed that 80 per cent of the boys, but only 61 per cent of the girls, owned digital watches. Another item in which a very substantial change was recorded was that concerning the mileometer of a car (see page 30 above). In 1984, 63 per cent of boys and 39 per cent of girls were successful at the time, compared with 54 per cent of boys and 42 per cent of girls in 1974. One might speculate that this change might reflect an increase in car ownership among the parents of young children, together with an increasing stereotyping of technological matters as a male domain.

Thus, it would seem that changes in both curriculum and technology are reflected in children's performance in mathematics at the age of 11. Continuing work will be needed to ensure that girls are not further disadvantages by changes that are bound to occur in the mathematics cirriculum due to the increasing use of technology in our society.

REFERENCES

Assessment of Performance Unit (1980) *Mathematical Development*, Primary Survey Report No. 1. London: HMSO.

Assessment of Performance Unit (1981) *Mathematical Development*, Primary Survey Report No. 2. London: HMSO.

Assessment of Performance Unit (1982) *Mathematical Development*, Primary Survey Report No. 3. London: HMSO.

Assessment of Performance Unit (1985) *A Review of Monitoring in Mathematics: 1978 to 1982*. London: HMSO.

Cockcroft, W.H. (Chair) (1982) *Mathematics Counts*, Report of the Committee of Inquiry into the Teaching of Mathematics in Schools. London: HMSO.

Eddowes, M. (1983) *Humble Pi: the Mathematics Education of Girls*. York: Longman, for the Schools Council.

Maccoby, E.E. and Jacklin, C.N. (1975) *The Psychology of Sex Differences*. Oxford: Oxford University Press.

Shuard, H.B. (In preparation) *Mathematics and the 10-year-old, plus 10*.

Ward, M. (1979) *Mathematics and the Ten-year-old*. London: Evans/Methuen.

2

*Attitudes and Sex Differences—Some APU Findings**

LYNN JOFFE and DEREK FOXMAN

ATTITUDES TO MATHEMATICS

The Assessment of Performance Unit (APU) has been concerned with the collection of information about pupils' attitudes to mathematics for two main reasons:

(a) because the thoughts and feelings of pupils toward the activities they engage in at school are an important feature of their learning; and
(b) because a positive approach to any school subject is an educational goal in itself, and data concerning any factors which may influence such an approach (attainments, for example) are considered important.

Using a written questionnaire, a wide range of data has been collected about opinions of 11- and 15-year-olds concerning mathematics and the relationship between attitudes and performance in written and practical tests.

From pupils' comments, it appears that there are at least two kinds of attitudes being measured in the questionnaire and quantitative data have been gathered on both of them. The first kind seems to be a broad, overall opinion; the type that may be elicited if one asks, 'What do you think about maths?' This aspect is reflected in pupils'

*This chapter first appeared in *Mathematics in Schools*, September 1984. The material contained in this chapter is Crown copyright, reproduced with the permission of the Controller of Her Majesty's Stationery Office.

ratings of the strength of their agreement with statements about mathematics. It seems likely that this general attitude is the one that will be most influential in older pupils' choices of mathematics and mathematics-related courses.

The second type is specific to a particular topic or item. Although pupils state preferences and give differential ratings of these, it seems unlikely that any one topic would be solely responsible for the pupils' general feelings towards mathematics. (Arithmetic might be an exception, as will be seen later.) It is more likely that success or failure might be rooted initially in one topic, which could trigger a positive or negative chain reaction, that might eventually influence the general attitude.

Some of the APU findings reveal that the relationship between pupils' perceived difficulty, usefulness and enjoyment of mathematics and their written test performance is weak. This does not mean that these factors are not important; they may have a broader influence on pupils' persistence when approaching mathematical material and may determine how much mathematics they choose to study.

Because the survey findings are so wide-ranging, it is not possible to detail them all in a limited space. This being the case, discussion in this article will concentrate on two features; (a) pupils' comments about mathematics generally; and (b) the presentation of some important aspects of both attitudes and performance which appear to be different for boys and girls respectively.

Pupils' comments give depth to the quantitative data collected through the structured parts of the questionnaire. They also give insights into a number of general issues that may be of interest to maths teachers.

Generally 11-year-olds' comments are less informative than those of 15-year-olds, partly because the younger pupils are less articulate, and presumably because they have had less exposure to mathematics and therefore do not have as strongly formed opinions about the subject as do 15-year-olds. For this reason, all comments mentioned below have been made by 15-year-olds, unless otherwise stated (G or B beside each statement indicates if it was made by a girl or boy).

There are a number of recurring themes which emerge from the comments. These will be presented below under appropriate summary headings.

What is maths?

Pupils tend to have strong opinions about what constitutes 'maths' and what does not. Questions that involve everyday experience (for example, reading a timetable or timing a recipe) are often not regarded as maths; a number of pupils make comments to the effect that:

*This is not maths, it's common sense.**

However, when items become more difficult, especially where calculation is involved, pupils refer to 'maths'. For example, in attempting to subtract decimals, many pupils made reference to the fact that:

I'm not much good at maths.

Their references are often to arithmetic specifically, though these feelings appear to colour much of pupils' attitudes to mathematics in general.

Maths and arithmetic

Often pupils use the terms 'maths' and 'arithmetic' synonymously. This appears to reflect common usage (in conversation, people often say things like '*I'm not much good at maths*', when they mean arithmetic). When referring to the items involving calculation, for example, many pupils said things like:

I know what to do, but my maths lets me down.

Mathematics as an emotive subject

It is evident from the comments that mathematics is a very emotive subject and that feelings about it often run high, as can be seen below. Strong, usually negative feelings are often engendered by the mere mention of the term 'maths'.

G *Some boys and girls do really bad in maths. It is not because they are stupid though . . . I myself am very bad at maths . . . you know (*that people are bad at maths*) by their attitude to maths . . . boys and girls panic or run out of time . . .*

Just because people are not good at maths it doesn't mean to say they are stupid. They could be really good at English. Somebody may be really terrible at English and ace at maths.

---

*Throughout this chapter, statements that appear in italics are pupils' comments, taken directly from survey scripts. No attempt has been made to correct spelling and/or grammar.

G *I think that most of the sums we do in maths are absolute rubbish and an entire waste of time. You will only need basic arithmetic anyway so why waste time doing something pointless when you could be studying for another subject.*

We did this type of maths once but our teacher tries to cram everything together and I don't understand. My maths does NOT deserve to be put in such a high class . . . I wish I could understand maths more clearly.

G *I don't like maths. I don't suppose anyone else feels the way I do.*

Maths as a useful subject

Utility is mentioned frequently in comments, and pupils often relate what they find useful from the mathematics curriculum to practical situations. Pupils also mention things that will be useful in the future.

G *I think maths is an* ***Important*** *subject because you need to know maths in everyday situations like addition, mult; subtractions, Division and money (and in jobs). Even if you are not very good at maths I think it is best to try and do the things that you don't understand as best you can.*

G *Although maths is not my favourite subject, I am glad I have learn't about it and can adapt some of the daily routine to it.*

G *I think maths is the most important subject in my school (apart from English).*

B *There should in my view be more about everyday maths in school, such as stocks and shares and calculating income tax.* (Perhaps not a representative view!)

There is some doubt expressed about the usefulness and relevance of some topics; many pupils mentioned that they do not see the point of much of the mathematics they do:

G *Some maths topics should not be taught because most people will not ever again use the problems in later life. But some should be taught and the important ones should be taught very well, so every child understands. I think it is stupid having so many topics. You should just concentrate on the important ones.*

From the ratings on the more structured parts of the questionnaire, it is clear that pupils of both age groups think that mathematics is a useful subject and, though they may have reservations about particular aspects of it, they recognise that they will need some knowledge of it both at school and in future life. Knowing that mathematics will be useful appears to be more important in influencing participation in mathematics than whether it is seen as easy or difficult or as enjoyable or not.

At age 15, boys rate mathematics as more useful than girls do.

Difficulty

Pupils mention the difficulties they have in mathematics in relation to particular topics and to the subject as a whole, as can be seen below:

> G *Things like this won't sink in to my brain.* (volume)

> B *I find that under exam condition or under pressure maths is a hard subject to do. One thing I am always doing is forgetting the formula to a question. Many people I know do this.*

At age 11, pupils' ratings of how difficult they find mathematics are related to how much they enjoy the subject, however, at age 15, these two factors are not linked—statistically, at least. For all pupils, the association between the perceived difficulty of the subject and written test performance is stronger than is the case for perceived utility or enjoyment.

In the older age group, ratings of difficulty are the greatest source of sex differences; girls indicate that they think mathematics is far more difficult than boys do.

Enjoyment

Amongst younger pupils, the enjoyment of mathematics seems to be more important in influencing opinions than is the case for 15-year-olds. In both age groups, though, the relationship between enjoyment and performance in written tests is weak. This does not mean that enjoyment is not important, since it could affect how much mathematics pupils choose to study and, as has been mentioned, enjoyment of the subject should be an educational goal in itself. It seems though that pupils take a pragmatic approach; if mathematics is seen as useful, they will do it, whether they enjoy it or not. It is gratifying though to find pupils saying:

> B *Maths is my favourite subject.* (11-year-old)

Confidence

An interesting finding is the frequent mismatch between performance and comments. Here, pupils who have been successful on an item still express doubt about their performance in general or attribute their success to luck. This appears to be more frequent in the case of girls:

> G *I cannot usually do these types of sums, but I can do this one cause it's easy.* (volume)

> G *I do not normally like them, but this one I do.*
> (fractions, 11-year-old)

> B *I always have problems with the decimal place, but not on this one it's easy.*
> (subtracting decimals)

> G *It's easy. I must be wrong.*
> (sets and Venn diagrams)

Attainment groups

The perceptions of pupils in different maths sets and classes (as mentioned by themselves) are illustrated by the following:
A 15-year-old girl who completed most of the written items in the questionnaire correctly, wrote:

> G *I find maths quite difficult, although I am a top stream candidate. However, sometimes I get really frustrated that I cannot understand my maths and sometimes feel that more time should be spent on maths in school. I feel it is one of the hardest subjects to understand.*

At the other extreme, a 15-year-old boy who got few items right, remarked:

> B *As you may have seen, we don't do a lot of hard maths in our group, perhaps it is because it's the bottom group, but I think that we should learn the things that we did not do.*

Another pupil who was successful on over half of the items, said:

> B *I think that people like myself would do better if we had more time to understand the work we do and then we would be able to get better marks and also have a better knowledge of maths.*

The sentiment of the latter remark was echoed frequently, particularly in the category below.

Pace of work, teachers and teaching

Comments about pace of work, teachers and teaching are common. They often indicate that pupils feel that their teachers do not recognise their difficulties. In response to a request to mention anything important about mathematics, not covered in the bulk of the questionnaire, one pupil said:

> B *Should always be plenty of examples for boys and girls to copy down. The teacher should always explain thoroughly. Should always help if pupil doesn't get the question. The teacher should always be patient with a pupil who can't get what the teacher is trying to explain.*

Others said:

> B *Let the teachers explain it more easily so you know what you are doing.*

> G *Maths should be taught at a slower pace than it is because this does not give people enough time to understand it properly. And the teachers should be more understanding.*

> G *I think that teachers tend to do the simple work not fully enough so that there are more problems when it comes to doing more complicated subjects.*

> B *. . . My teacher doesn't show us how to do them the easy way, but teaches us how to do it the baby way, which we find very complicated and sick, he thinks we are baby's!*

What turns pupils off

Insights were gained into what makes pupils 'turn off' aspects of mathematics. Although specific instances are given, there is evidence to suggest similar widespread attitudes to many topics.

In response to a word problem about speed and velocity, one pupil said:

> B *Too many words to describe the question made it seem difficult.*

One 11-year-old said of an item concerned with money:

> B *I don't like it because they could put it as an ordinary sum instead of words.*

Faced with calculating the volume of a box, given a diagram and measurements, another pupil remarked:

> G *Although I have not done these, but just looking at them I don't really want to know about volumes.*

Comments like this serve to emphasise that the visual impact of mathematical material should not be underestimated.

Specificity of knowledge

For many items there were comments about the fact that the topic had been done but was now forgotten. Pupils appear to have pockets of knowledge which are specific to the time they 'did it in class' or to a particular situation. Knowledge is not generalised in these cases:

B *I have done it before but forgotten how to do it.*
(volume, 11-year-old)

G *We did these in class some time ago but I can't remember how to do them.*
(Venn diagrams)

B *We did these in Physics but I've forgotten where to start.*
(velocity)

SEX DIFFERENCES

Sex differences in attitudes

Although pupils of both ages, in all surveys, have been asked to rate the extent of their agreement with statements about whether boys and girls perform differently in mathematics, a semi-structured section, designed to gather more information, was included for the first time in the 1982 survey of 15-year-olds. Pupils were asked in detail about sex differences in mathematical ability: a large proportion of both boys and girls said that there were no differences, though some of these did go on to detail differences, along with those who answered 'yes'.

Amongst those who thought there were differences, opinion was divided as to which sex was superior. In favour of girls, the following comments were made:

B *The girls tend to get better results, because they seem to be able to concentrate more.*

G *Girls do best. Most of the boys mess around.*

B *When questions are asked girls always seem to know the answer.*

Where differences were seen in favour of boys, the following statements were made:

G *Most boys do maths but not many girls do maths.*

B *In some cases I think that boys tend to have a more logical and clear view of maths.*

B *Boys have more ambition in life therefore they work harder.*

G *I think boys learn the more basic things and use computers etc because they've got a better chance of getting a job to do with maths.*

Some pupils strongly refuted the suggestion of sex differences in mathematics:

No, that's rubbish.

Others were not as sure:

G *Boys seem to take more interest in maths — but girls do quite well.*

One boy, who was ambivalent about this question about sex differences (though not about sexist language), said:

B *It is said that boys have more ability in such subjects, I don't know what the differences are, but I do my bird's homework.*

Other pupils' comments bear out findings from other studies (see Spender 1981 and Spender and Sarah 1981):

G *Teachers especially men seem to pay more attention to boys in the same subjects.*

G *Boys may need more maths to do computer studies etc., these opportunities are not given to a girl.*

B *It's often thought that boys are better than girls, but I don't think so. Often girls give up when hard reasoning is involved.*

B *If some girls do not know something, they keep quiet about it.*

B *When a boy is asked what a question is he ansers it straightaway.*

B *... when you ask a girl a maths question she takes about 5 minutes to answer it... Boys do more complicated jobs than girls a boy learn easier aswell.*

B *The girls don't need maths as a housewife but men need them to support the house.*

B *Boys are more inteligent than most other girls.*

G *More bosses of shops, people working in banks are men.*

G *Boys are more apt to use Maths in Engineering.*

G *Both my brothers did Maths for A level which I will not do and nor did my sister.* (from an 'O' level candidate)

G *At girl schools, people tend to go more for the Arts and people who can do Science at O level think they cannot do them for A level.*

Some general aspects of sex differences have emerged consistently through the surveys:

1. When asked to rate statements and indicate the perceived difficulty and usefulness of mathematical topics and items, girls tend to make more moderate assessments; they use the extremely positive and extremely negative positions on the rating scales far less than boys do.
2. Girls express greater uncertainty about their mathematical performance. Boys express a greater expectation of success.
3. Boys overrate their performance in mathematics in relation to written test results; they do not do as well as they expect to. Girls underrate their performance and do better on tests than they expect.

For example, for the item in Figure 2.1 48 per cent of boys and 42 per cent of girls gave the correct answer.

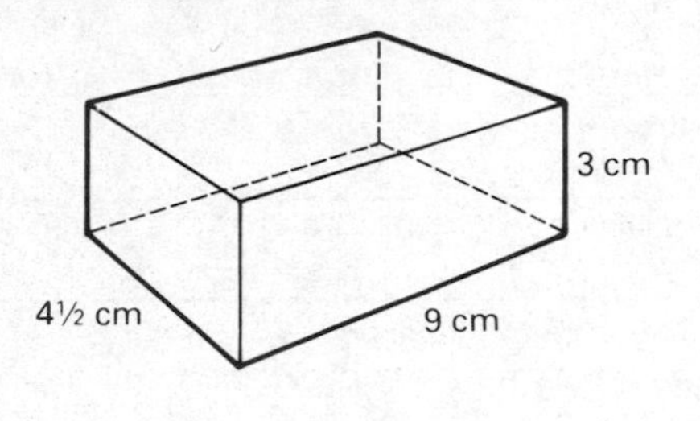

What is the volume of this rectangular block? __________ cm^3

Figure 2.1

When asked to rate how easy/difficult it was, 51 per cent of boys and 38 per cent of girls said it was easy—more boys and fewer girls gave the correct answer.

In one survey, pupils were not required to complete the item; they were just asked to look at it and say how easy/difficult they thought it was. In this case, both groups underestimated the difficulty, though boys did so more than girls; 58 per cent of boys and 51 per cent of girls said it was easy.

Sex differences in test performance

We have mentioned briefly that boys and girls have differing attitudes towards some mathematical topics. These differences are echoed in the test performances of the two groups of pupils.

The surveys have shown three main features in relation to differences in performance between the sexes:

1. The profile of performance of the sexes across some topics remains relatively constant between ages 11 and 15.

   There are significant differences in mean scores of boys and girls on a number of different topics but, overall, boys are best, relative to girls, in the applied and practical areas—measures and rate and ratio—while girls do best relative to boys in computation with whole numbers and decimals and some aspects of algebra. By age 15, girls are marginally behind the boys in the latter topics but more substantially so in the former. This is illustrated in Table 2.1 which shows boys' minus girls' mean scores in measures and computation with whole numbers, at the two ages.

Table 2.1

			Boys' minus girls' mean scores at ages:		Scores at age 15 minus scores at age 11
			11	15	(b)
Topic:	Measures		+4%	+7%	+3%
	Computation		−2%	+1%	+3%
Measures scores minus computation scores		(a)	+6%	+6%	

From the table it can be seen that:

(a) the relative difference in the two topic scores is similar at both ages (6 per cent);
(b) the change in boys' scores relative to those of girls is similar for both topics (3 per cent);
(c) the differences which already exist at age 11 (a) are twice as great as those which take place between ages 11 and 15 (b).

Although it is recognised that there are differences within these topic areas at both ages, the between topic scores appear to be of particular note.

When the results are examined more closely, it appears that:

2. The differences in performance are minimal in most topic areas—except in the top attainment bands.

   Nearly all the differences in performance between boys and girls at both ages are accounted for by the top 10 to 20 per cent of attainers in most topics (top 30 per cent in measures and applications). The proportion of boys to girls among 15-year-old pupils obtaining the highest 10 per cent of scores on APU concepts and skills tests is 61 to 39 per cent. Among the highest scoring 10 per cent of 11-year-olds on the comparable APU tests, 58 per cent are boys and 42 per cent are girls.

The greater proportion of boys than girls in the top 10 and 20 per cent bands is not matched by a corresponding imbalance of more girls than boys in the bottom 10 and 20 per cent bands. Indeed, at age 11, there is a tendency for more boys than girls to be in the bottom 10 per cent in most topics.

3. The main differences in performance are already established by age 11.

We have noted that, by the age of 11, boys express more confidence than girls in their chances of success in tackling questions on measures topics. Boys are also observed to be more confident in their approach to measurement tasks, particularly in the practical tests. This may be puzzling given that other mathematics tests given to 10- or 11-year-olds often do not produce differences in the performance of boys and girls respectively (ILEA 1983). This is probably because in these tests, items favouring boys are balanced by those favouring girls.

In contrast, the results of the APU tests and attitude questionnaires suggest that the pattern of differences between boys and girls, both in their attitudes to mathematics and in performance, has been firmly established by the time the pupils leave their primary schools. These findings are of particular interest if taken in conjunction with those relating to attainment bands. We have found that the top 20 per cent at age 11 are performing as well as middle attainers (40–60 percentiles) and, in some cases, upper-middle attainers (60–80 percentiles) at age 15. A pupil in the top band at age 11—and the majority are boys—must therefore have a very good chance of being in the top band at age 15.

Since there is no known innate reason for differential performance in and attitudes to mathematics of boys and girls (Spender 1981, Spender and Sarah 1981 and Whyte 1983), one has to go beyond the scope of APU data for explanations of their measured findings. For example, HMI in their 1973 survey of curricular differences of boys and girls in mixed and single-sex schools (DES 1975) noted that, even in primary school, boys engage more than girls in using tools, construction, three-dimensional modelling and measurement. Since social reasons are often given as the basis of sex differences in attainment, it may be that the stereotyping of activities as masculine or feminine is reinforced in schools as well as in society in general.

One final point is important. So far only a few analyses of problem solving by sex have been carried out. Those which have been completed in this area of mathematics have not revealed any significant differences in performance between boys and girls.

FUTURE WORK AND PUBLICATIONS

A comprehensive account of the results of the phase of annual APU mathematics surveys (APU 1985) has been published. The results are wide-ranging and the document is correspondingly weighty.

In order to make the findings more accessible to specialist audiences, shorter booklets will be written, each discussing particular aspects of the surveys in more detail than has been possible in published reports. For example, documents dealing with the assessment framework, problem solving and decimals are in preparation.

The next mathematics survey will take place in 1987. In the meantime, more in-depth analysis of survey data will be undertaken, as well as investigations into how pupils communicate mathematical ideas, patterns of individual performance and background variables.

REFERENCES

APU (1980), (1981), 1982) *Mathematical Development*, Primary survey reports Nos 1, 2 and 3. London: HMSO.

APU (1980), (1981), (1982) *Mathematical Development*, Secondary survey reports Nos 1, 2 and 3. London: HMSO.

APU (1985) A Review of Monitoring in Mathematics, 1978 to 1982. London: HMSO.

Department of Education and Science (DES) (1975) *Curricular Differences for Boys and Girls in Mixed and Single Sex Schools*. Education survey 21. London: HMSO.

Inner London Education Authority (ILEA) (1983), *Race, Sex and Class 1, Achievement in Schools*.

Spender, D. (1982) *Invisible Women*. London: Writers and Readers.

Spender, D. and Sarah, E. (eds) (1980) *Learning to Lose*. London: Women's Press.

Whyte, J. (1983) *Beyond the Wendy House. Sex Role Stereotyping in Primary Schools*. Harlow: Longman/Schools Council.

3

Sidetracked? A Look at the Careers Advice Given to Three Fifth-Form Girls

DIANNE

Occupational interest as expressed at the start of the fifth form—laboratory assistant or radiography.
Shortly after the start of the fifth form Dianne changed her mind and decided to concentrate on engineering.
Educational qualifications—six GCE 'O' level passes including mathematics, physics and chemistry, plus three CSE examination passes.

There are no careers 'lessons' at Dianne's school and the only career-related activities which had taken place in the fifth year were visits by representatives of the army and the police. Attendance on the part of the pupils on these two occasions was a matter of individual choice.

There was a teacher at this school who had the responsibility of providing advice about careers but Dianne had never made an appointment to see him.

> *However, you can do so if you want to. You have to make an appointment. I was applying for jobs on my own so an appointment didn't seem worth it.*

---

*This chapter originally appeared in *Sidetracked? A look at the careers advice given to fifthform girls* published by the Equal Opportunities Commission in 1981. Reproduced with permission.

Dianne was interviewed once by a careers officer in her fifth year. She attended this interview with the idea of becoming a radiographer or a laboratory assistant, the possibility of engineering as a career had not yet occurred to her.

During the interview Dianne felt that the officer was really pushing the idea of full-time further education (in order to train as a radiographer 'A' levels would have been essential):

> *I feel that the careers officer could have been more helpful. All he did was to spend all the time talking about college. He said being a radiographer would be difficult to get and hard to do. He didn't even mention anything about laboratory work—no advice and no addresses to write to concerning vacancies.*

The officer did give Dianne the names of local hospitals to contact. No alternative occupations were suggested during the interview nor was the possibility of a non-traditional career explored.

It was some time after this interview that Dianne 'discovered' engineering when looking through the employment advertisements in the local paper. It appealed to her a great deal:

> *It seems the kind of job I would like.*

However she found that when she mentioned her interest in engineering to teachers at school she did not receive much encouragement or guidance.

> *The teachers just laughed at me*

At school Dianne had never been shown how to collect information relevant to her occupational interests, but on her own initiative she had written to a large number of firms to request information. She had approached several engineering firms enquiring about apprenticeships. Many of these firms did not reply. And although she had specifically requested information concerning engineering apprenticeships, in some cases she had been sent information and application forms for clerical employment.

It was a conscious decision on Dianne's part to leave school at the age of 16 and start her career at this stage. She had been told by the teachers at school that she had the potential for a university degree course, but she argued that by taking a job she could get further than going to college (and be paid):

> *I want a career very much—I am determined and motivated.*

As a result of one application for an engineering apprenticeship with a local firm, Dianne was asked to take a selection test, which she passed. She was subsequently interviewed by the firm's personnel officer, but was unhappy about the way the interview was conducted:

> *He asked how I would cope if I rose to the top of the firm or if I would be satisfied with a lower job . . . he made it clear that he didn't think I would get the job and that he didn't want me to get it. He said 'we have never had a girl here yet'. The atmosphere was very tense. He asked how I would feel working with men, he went on a lot about this...he kept plugging leadership potential.*

Dianne's interview with the personnel officer lasted for approximately twenty minutes. She commented on the fact that the previous applicant to be interviewed—a boy from the same school—had been given much more time, about forty-five minutes. This boy was offered an apprenticeship even though, as Dianne said with great conviction, his educational standard was below her own. At school the majority of subjects which he had studied were at CSE level. Dianne had already obtained a pass in mathematics at GCE 'O' level in the summer of her fourth year at school and the majority of subjects which she was studying were at 'O' level.

Dianne felt very strongly that the reason she was unsuccessful in her application for an apprenticeship with this firm was because she was a woman.

She is now working as a clerk for a group of accountants.

PAMELA

Occupational interest as expressed at the start of the fifth form—banking.

Educational qualifications seven GCE 'O' level passes including mathematics and English language.

Careers activities in Pamela's school began in the fourth-year and consisted of one lesson per week throughout the year. Pamela felt that this single weekly lesson was insufficient and would have welcomed additional lessons.

During the fifth year the school separated boys and girls for careers activities and as a result the careers discussed in Pamela's group were mainly the traditionally female occupations:

> *This year we have covered a lot of office jobs as it has been an all girls group. The boys have separate lessons where they have all the boys' subjects I suppose.*

Careers lessons were not available to all pupils in this school, it depended very much on the subjects chosen for study. Latin, for example, was regarded as an 'extra' subject and pupils following this course did not have careers lessons. Apparently this is not an uncommon situation.

The careers activities consisted in the main of viewing special television programmes designed to portray a variety of occupations. Pamela could only remember watching programmes which dealt with retail and clerical types of employment. She did not feel that these activities had assisted her in making her career choice since her ideas were already quite firm:

> *I have always wanted to work in a bank for as long as I can remember, so my ideas were not really formed out of these activities.*

Pamela had never discussed the possibility of non-traditional careers for girls in any of the activities she had attended.

In this school a careers teacher was available on an informal basis if pupils wished to approach him for advice. Pamela, although aware of this, had not contacted him.

Pamela had attended two interviews with a careers officer. She found the advice she received to be generally helpful. The officer gave her basic information, such as the entry qualifications required and addresses of local banks to contact. He suggested that she contact the banks quickly to request application forms so as to be among the first applicants for vacancies, saying that there was a good deal of competition among school leavers for employment in banks. Pamela was slightly discouraged by this.

During these interviews there does not seem to have been any discussion of the types of careers available within banks, nor of the Institute of Bankers qualification which is so important for career progression. The importance of obtaining day-release during working hours to enable employees to study for this professional qualification was not given any emphasis.

Pamela only gained further information about the type of work available in a bank when she attended an interview for employment. She was interviewed by a bank manager, as she recalled:

> *I had a maths test. The manager asked about my family and interests. He did most of the talking, telling me about what I would do (in the bank). He said that day-release (to attend a college of further education) was mainly for men, for those men who want to become managers. He said he discourages women (from going on day-release) because they tend to leave and have babies and break their career. Whatever I would learn would be from someone in the office, not going to the 'tech'.*

Pamela disagreed strongly with the manager's sentiments and felt that day-release would most certainly benefit her career.

I would have more chance of being promoted and it would improve my knowledge of the bank.

Pamela was offered a post and is now working at the bank. At the time we last interviewed her, it was not clear whether she would be successful in obtaining day-release.

MARIE

Occupational interest as expressed at the start of the fifth form—Mechanical and chemical engineering
Educational qualifications—five CSE examination passes including English language and physics.

Marie was very interested in engineering, particularly in chemical or mechanical engineering. She attended career talks and lessons from the fourth year onwards at school, usually for one hour per week throughout the year. The careers lessons mainly provided general information about job applications but Marie particularly remembered discussing banking, the forces, textiles, working on the buses, and engineering. These occupations were illustrated in TV careers programmes and Marie gained the impression that they were for both boys and girls.

It was pointed out that they (engineering firms) needed girls although it had been thought originally to be a boys' occupation.

The school provided the opportunity for individual counselling with a careers teacher, though on an irregular basis. Marie had taken advantage of this and explored the possibility of a non-traditional career with her careers teacher. This she found very helpful.

Marie was first interviewed by a careers officer in the fifth year and seems to have been particularly fortunate in the guidance offered to her. She had two types of interview. The first aimed to help pupils 'choose the right career', this was mainly a diagnostic service. The second was mainly concerned with providing information on the occupation chosen by the interviewee with some slight probing by the careers officer to confirm career choice.

The careers officer informed Marie that the entry requirements for engineering were passes at the GCE 'O' level (or equivalent) in mathematics, physics, chemistry, and English language. However, the majority of firms seem to recruit apprentices on the basis of selection tests administered before the GCE 'O' level and CSE examination results are known.

The careers officer gave her further information on the method of applying for jobs; he gave her the addresses of firms to contact, and two application forms for an apprenticeship with local firms. He also suggested alternative occupations in non-traditional areas, including marine, electrical, nuclear and electronic engineering whilst ultimately supporting Marie in her preference for either mechanical or chemical engineering.

Marie set about collecting a range of engineering job descriptions from a variety of sources. The information about the chemical industry gave her the impression that the most usual career route was through entry at degree level. Marie had already made up her mind to leave school immediately after the 'O' level examinations and so her final choice rested with mechanical engineering.

Marie has high aspirations:

> *I am aiming for a management career within an engineering setting. A career is very important to me and I am prepared to work hard to get as high as I can.*

She was enthusiastic about training, but felt that day-release opportunities for women were very often limited.

> *I think that if girls are really determined to do engineering, for example, they should have the same opportunities as boys. Shorthand and typing is not my idea of a career, and that is the kind of day-release that many women get.*

From the start Marie anticipated some difficulty in finding an engineering job:

> *When they (the employers) want girls in engineering, they require a higher standard of qualifications than for boys. A lot of boys at my school have got jobs in engineering although their educational standard is lower than mine.*

A local employer responded to Marie's application for an apprenticeship in engineering by inviting her to take a selection test. Marie passed the test and subsequently was asked to attend an interview. The firm had two vacancies available, one a two-year apprenticeship in 'store keeping and accounts', the other a four-year apprenticeship in 'plant fitting'. Marie was very interested in the second vacancy as the first 'was going away from what I wanted to do' and in her view offered 'very little chance for promotion'.

She described the interview thus:

> *I think the personnel officer was a little embarrassed when he interviewed me. He said that they very rarely had girl applicants and that I was the first girl he had interviewed. He said it might be better for me to*

> *go into store keeping as I might get teased a lot on the shop floor and would have to cope with bad language. But he also said my presence on the shop floor might have the effect of calming down the men and that would be a good thing.*

Marie was offered the job as apprentice store keeper. She was aware that the training given was limited and that there were few career prospects. Despite this she accepted the job in preference to becoming unemployed, conscious that she was moving away from her real interests and aspirations.

4

Some Thoughts on the Power of Mathematics: What is the Problem?

DALE SPENDER

The celebrated anthropologist, Margaret Mead, once stated that the world over, men may weave or dress dolls or hunt humming birds, and whenever such activities are seen as appropriate for men, the whole society—women and men alike—accords them great significance and prestige. Yet, she asserted, that when the very same activities are undertaken by women the whole society deems them to be insignificant, and even to constitute drudgery.

This principle, that the status of the sex can determine the status and desirability of an activity—is one which has relevance outside anthropological circles. It is a principle which has its uses in educational theory and practice. It is a principle which can serve to remind educational researchers that when a high-status subject is seen to be the preserve of a high-status group (and vice versa, when a low-status subject is primarily the province of a low-status group) then it could be that status and its links with power and dominance is the predictor of performance, attainment and attitude.

For example, only if issues of power and dominance are kept out of consideration would it be possible to describe the supposedly superior sex performing in a supposedly superior manner in a supposedly superior subject as a *problem* of girls and mathematics.

Anyone who has observed boys defending their mathematical territory (and computer time) from the entry of girls, would find it difficult if not confusing to be asked to explain such behaviour in terms of either the inherent abilities of girls or the inherent properties of mathematics. For when boys clearly understand and readily acknowledge that mathematics is a subject which can open many career doors—and then roundly declare that girls cannot do mathematics—it is not necessary to be a particularly perceptive person to plumb the reasons.

When girls who demonstrate any inclination or aptitude for the high-status mathematics are 'warned off' by the boys and directed towards their proper places in low status subjects (which are unlikely to open any career doors), it would be a 'short sighted' researcher who attempted to explain the predominance of girls in domestic science in terms of the limitations of female spatial ability.

Of course, the problem could be that girls *allow* themselves to be 'seen off' from the subject of mathematics. But again, to make this a failure of the girls, is to engage in scapegoating: it is to blame the group without power for not having power. It is to deny the complexity of our socialisation processes and to omit considerations of the asymmetrical relationships of the sexes in our society.

Girls are reared in a world where females are encouraged to be emotionally and financially dependent upon men: according to the UN men own more than 99 per cent of the world's wealth and are increasing their resources every year. And this makes women (and their children) a dependent group, despite individual exceptions. What needs to be acknowledged is that any dependent group is often obliged to find favour with those on whom they depend: women are no exception. And if the men on whom women depend want to be seen as superior at weaving, or dressing dolls, or doing mathematics, it is not unlikely that women will 'withdraw' from the competition and leave the men to parade their special and superior skills in such areas of endeavour.

That the problem of the male monopoly of mathematics is not necessarily going to be solved by reference to girls, or to mathematics, but with understanding of territorial rights, advancement and power, is well illustrated in the case of classical languages. Last century, when it was classical languages which opened career doors, it was widely established by reputable men, that girls could not do languages. But now that the focus has shifted and languages are not the testing ground for hierarchies and success, girls have been found to be very good at the low-status subject of languages. It is mathematics that they find difficult now. And presumably if next century most power is still concentrated in the hands of men, and child rearing is decreed as the crucial determinant for career advancement, it will soon be demonstrated that girls have no aptitude for child-rearing practices.

And if the girls then are as circumscribed by dependence as the girls now, they will quickly readjust their sights and look for more appropriate occupations than child rearing. Of course, child rearing would soon become a high-status activity (as well it should be) and no doubt demands would start to be made on behalf of women for equal access to child-rearing education and occupations. There would probably be a spate of research responses which could readily establish that it was either women's hormones or women's brains (both having served as excuses for more than a century) which were responsible for their deficiences as child rearers. Then special education programmes could be devised to help women overcome their inadequacies—and to play an equal role in the high-status activity of child rearing. Just as we can mount special programmes to help girls take an equal place alongside boys in the high-status subject of mathematics.

Such a scenario is not absurd—although it is unlikely. But it does help to point to some of the limitations of the frame of reference in which much of the debate on girls and mathematics takes place. While many students may be good or bad at mathematics, while many students may like or dislike mathematics, when mathematics is seen as a boy's subject (and languages are seen as a girl's), then its neither to girls nor to mathematics that we need to look for explanations of the dynamics. 'Mathematics' is part of the 99 per cent of the world's resources owned by men and they guard it well.

5

Girls and Mathematics: The Negative Implications of Success

ROSALINDE SCOTT-HODGETTS

INTRODUCTION

Until recently, nobody had seriously questioned the belief that, certainly by the age of sixteen, boys' performance in mathematics is in general superior to that of girls. (Note that it is performance or attainment that we are talking about here, and not *ability*.) However, the recently published *Girls and Mathematics: From Primary to Secondary Schooling* offers a radically different view of the position of girls in the mathematics classroom (Walden and Walkerdine 1985); this publication is a revised and extended version of their original report of a three-year research project (Walden and Walkerdine 1983).

The work of Walkerdine and Walden has provided a wealth of insights into teacher–pupil relationships and other sociological and psychological aspects of mathematics classrooms. However, some of the conclusions which they draw are, by their own admission, contentious.

The research team directed by Valerie Walkerdine started with the hypothesis that there was a discontinuity in performance of girls between the ages of eleven and sixteen; overall, it was held, girls were performing as well as, if not better than, boys at the end of their primary schooling (APU 1980, Douglas 1964), and it was assumed that something was going seriously wrong for girls at the secondary level. However, having carried out a programme of testing and observation in primary and secondary schools, Walden and Walkerdine reach a different conclusion:

> What is suggested by the tests and other data is that there is far more continuity than we had imagined. However, it is a continuity of good rather than poor performance. Even in the fourth year secondary tests, girls are still doing well in comparison with the boys.

The samples tested by Walden and Walkerdine were very small (for example, the fourth-year tests analysed involved responses from only 97 pupils, all of whom attended the same school). However, their rejection of commonly held beliefs about differential performance is not based on their own data alone: in their revised report (Walden and Walkerdine 1985) they provide a theoretical rationale for critical reappraisal of interpretations of the results of large-scale surveys undertaken by the DES Assessment of Performance Unit (APU 1980, 1981, 1982), and of tests administered to 2300 pupils as part of the Schools Council Project *Mathematics and the Ten-year-old* (Ward 1979). Their argument is a powerful one, based on the valid criticism that commentators have failed to make a clear distinction between statistical and educational significance (a fault of which *this* writer has certainly been guilty).

The exposition of Walkerdine and Walden deserves a more detailed response than is possible here, but there are two relevant points which must be raised:

1. The fact that differences are statistically significant means that they *are* likely to be 'real', in the sense that such differences would probably be found if the total population were tested.
2. In their discussion of the 'triviality' of reported differences, Walden and Walkerdine, like some writers of whom they are critical, appear to make the assumption that the differences must be equally distributed over the whole female/male population! Such an interpretation of the available evidence would clearly be unreasonable; a more sensible conclusion to draw (there are acceptable variations) is that, whilst both girls and boys find mathematics difficult, in some areas of the curriculum fewer girls than boys achieve a very high level of attainment.

This disregard of distribution is also apparent in their analysis of their own research data:

> Despite the fact that, in the mock examinations which we have analysed in this chapter, girls performed overall better than boys, in four out of five tutor sets, hardly any girls at all were entered for O level . . . Our interpretation of the data suggests, conversely, that *despite* their success in the fourth year relative to boys, girls are not

> being entered for O level in anything like the same proportions as boys.

They go on to suggest that it is not girls' performance that is the problem, but teachers' interpretations of their performance, in that girls' performance is being evaluated in a pejorative way because what is understood as feminine behaviour is understood by teachers to be antithetical to real learning. This prejudice is, they suggest, so powerful that girls have to do *better* than boys to be entered for 'O' level. The problem with this argument is that they also state that the standard criterion for entry into the 'O' level set was a mark of 82.5 (three girls and seven boys were entered despite having lower marks). From their published data, then, it is clear that although on *average* the girls in their sample are doing well, there are fewer girls than boys achieving marks of 82.5 or above (in fact the ratio is approximately 1:2.6).

Having made the above criticisms, it should nevertheless be stressed that the work of Walkerdine and Walden makes some profoundly important points, both in terms of their own research and in relation to the way in which we interpret the work of others. It is quite clear, for example, that the teachers interviewed during their study *do* value 'success' in the classroom according to their perceptions of *how* such success was achieved, and this certainly requires explanation. But perhaps the most important issue they raise concerns the extent to which the 'problem of girls' performance' exists at all. It is important to dispose of the myth that all, or even a substantial majority of, girls are performing poorly in mathematics relative to boys. The argument which suggests that *no* educationally significant differences exist is, however, also unconvincing, because of the conflict which arises when intelligent interpretations of existing research data are considered.

It is the purpose of this article to present a hypothesis which is believed to be consistent with the data published by Walkerdine and Walden; it answers their challenge to provide an explanation to account for girls' success as well as their failure, but allows a modified interpretation of their observations, more compatible with evidence available from other studies.

REPORTED GENDER DIFFERENCES IN MATHEMATICAL ATTAINMENT

It is the belief of the writer that the study of patterns of performance is important in seeking to understand the position of girls in relation to mathematics. In undertaking such a study, however, it is important to maintain an awareness that the mean differences referred to are, on the whole, very small, even when they are labelled 'significant', because it is *statistical* significance which is implied.

One of the most important developments in recent years has resulted from analysis of the APU tests, which have allowed us to look at the relative performance of girls and boys separately in various topics within the mathematics curriculum. Lynn Joffe, in this volume, reports on the differences between girls and boys, both in performance in core aspects of the curriculum and in attitudes to the subject, which emerge in the APU surveys by the age of eleven, and could be seen as prefiguring wider differences between girls and boys at the age of fifteen (Chapter 2).

Hilary Shuard presents more structured information about the rank order of differences between boys' and girls' scores on different mathematical topics; this ordering remains much the same over the three surveys of eleven-year-olds undertaken by the APU (Chapter 1):

Combined rank order of differences between boys' and girls' performance on various mathematical topics at age eleven

(a) application of number;
(b) length, area, volume, capacity;
(c) rate and ratio;
(d) concepts of fractions and decimals;
(e) computation with fractions;
(f) money, time, mass, temperature;
(g) concepts: whole numbers;
(h) symmetry, transformation, co-ordinates;
(i) probability and data representation;
(j) lines, angles, shapes;
(k) sets and relations;
(l) generalised arithmetic;
(m) computation: whole numbers and decimals.

Those topics at which boys excel most are at the top of the list, and those at which girls have the greatest advantage over boys are at the bottom of the list. It should be noted that only in the last two topics did girls perform better than boys in all three surveys. In each survey up to five of the differences were significant, and the pattern of differences, as well as being largely consistent over the three surveys, is broadly in line with the findings of research in the United States (Armstrong 1980).

An analysis of test results from the study *Mathematics and the Ten-year-old* is offered as additional evidence of an emerging pattern of differential performance.

In over a quarter of the test questions there was a significant difference between the performance of girls and boys; in eleven questions girls did significantly better, whilst boys were significantly more successful in fourteen questions:

Differences in girls' and boys' performance in tests for the Schools Council Project, 'Mathematics and the Ten-year-old'.

Girls were significantly better at:

(a) six questions concerned with computational skills with whole numbers and money;
(b) three questions set in purely in verbal terms involving naming geometric shapes and making a deduction from given (non-numerical) information;
(c) one question in which a pattern had to be followed;
(d) one on the relationship between two sums of money.

Boys were significantly more successful at:

(a) four questions testing understanding of place value;
(b) other items concerned with measurement, spatial visualisation, problem solving, reversing an operation.

Certain similarities with the APU test results are apparent, but of particular interest are the responses of teachers (87) and head teachers (38) of these pupils to a questionnaire asking them to rank the test questions in order of importance. The rationale for the setting of the questionnaire was that the feedback would give some indication of the priority which teachers gave to the teaching of particular topics. It was found that the questions at which girls did significantly better were ranked higher than those at which boys did significantly better. Shuard suggests that it is perhaps even possible 'that pupils who pay attention to the teacher's traditional emphases in primary mathematics give themselves a positive disadvantage for future success in mathematics' (Chapter 1).

The following extracts from an interview with a ten-year-old describing the situation in her classroom serves as an exemplar of the recognition of a teacher's priorities by her pupils, and suggests a difference in response between boys and girls within the class. The interview was conducted by the writer for inclusion in a presentation for PGCE students:

S = Susan *I* = Interviewer

S I think girls do better (at maths) because boys—if we have to do, say, three pages a day, they just scribble it all down; they don't care. But girls do care, and they don't care if they finish their work or not as long as it's tidy.

I Sorry—you say girls don't care whether they finish the work?

S Yes, and so does(n't) our teacher. She doesn't care whether we finish our work or not so long as it is tidy . . . and we've done as much as we can.

I Do you ever do any practical maths?

S Not usually.

I Sometimes?

S Sometimes; we have got geoboards and things like that, but the geoboards are all I've ever used in the third year, because Mrs. Browning* doesn't care really.

I Doesn't care about what?

S Volume and everything. When you do volume she says 'Oh, leave it,' or 'You can do the bit you can do without water.' And in the second year we were allowed to use water and we were allowed to to go out and make charts and draw on them, and draw chalk circles.

I Ah. So you haven't been doing so much of that. O.K. So think back to last year then, when you were doing lots of practical work in maths. Was there any difference between boys and girls there?

S No, I don't think so.

I You think that girls and boys were equally good at that sort of work?

S Yeh, but I think in the third year the boys have dropped back because they are so bored that they don't go outside and play with water and everything for their volume and all that.

I Oh. That's interesting. So you think it's because of the type of maths they are doing that they are less interested? And the girls do it anyway . . . Why do you think that is?

S Because I think the boys had a better time when we were outside with the water and everything, because they mucked around and spilt water and chased people with worms, and now all they can do is sit in class and play with toys.

I Right, but why do you think the girls are different? Why do you think the girls keep doing the work and the boys don't?

S 'Cause I think the girls know that one of these days we're going to get work like that, as soon as we can please our teacher, and make sure that she likes everybody's work.

I . . . What sort of maths do you think your teacher feels is important?

---

*This name has been changed.

> *S* Um. The adding, and times, things like that. And she likes us to write it all out, the sum, and not think in our head and then write it: like 3 chickens—if 3 farmers had 9 chickens, how many would they have each. Not 3 chickens—she'd want us to write it out: 9 divided by 3 equals 3 chickens. We've got to write it out. Neatly.

During the course of the interview, Susan was asked to think of two pupils, a boy and a girl, who she felt were good at maths, and to say what it was about them that made them good. In both cases she attributed their (perceived) success to hard work. The boy, who was judged best in the class at everything except art, was assumed to work hard, 'probably from the age of five, learning his tables'. The girl was noticeably well behaved: 'She doesn't care if someone's talking or anything; she just ignores them and just carries on with her work'. And, of course, they were both neat and tidy!

It is not suggested that the situation described in this interview is typical (it is certainly to be hoped that this is not the case!). However, the evidence about teachers' priorities demonstrates that Mrs Browning is not alone in thinking that the most important elements in mathematics are 'adding and times and things like that'.

In this particular classroom, the boys are opting out of the boring work but the girls are carrying on conscientiously. Why?

There is evidence to support the view that girls are generally more concerned about adult approval (e.g. Maccoby and Jacklin 1975). Also important are adults' differing expectations of girls and boys (see Walkerdine and Walden 1982, 1983; Walkerdine 1983). Walkerdine quotes from the Plowden report:

> The high spirited mischievous boy is traditionally regarded with affectionate tolerance. 'Boys will be boys' . . . a boy who never gets up to mischief, it is suggested, is not a proper boy.

She asks why the naughtiness of boys is seen as constructive whilst that of girls is simply naughty. Her interviews with primary school girls revealed that they were disdainful of badly behaved boys who break the rules and don't work hard. Walkerdine makes various interesting and plausible suggestions to explain these behaviour differences, but the underlying causes will not be pursued here; the premise will be accepted that girls, for whatever reasons, feel compelled to be good.

To recap: when the balance of items in a test reflects the priorities historically found in primary school classrooms, girls' overall mathematical attainment is as good as that of boys; girls are expected to be well behaved, and are interested in pleasing their teachers; these teachers believe that a particular range of elements should take priority in mathematics education; girls perform most competently in these 'priority' areas of the mathematics curriculum.

The hypothesis being put forward in this article is founded on the belief that some girls are indeed at a disadvantage in mathematics by the end of their primary schooling, but that this is not solely the result of the affective factors which encourage them to concentrate their efforts on particular areas of curriculum content. It is suggested that some pupils (both boys and girls, but more girls than boys) may be adhering to a particular set of strategies which have led to their success in mathematics, but which, when used exclusively, have negative implications for these pupils' mathematical development.

THE SERIALIST/HOLIST DICHOTEMY

Fennema maintains that, 'since mathematics is a cognitive endeavour, the logical place to begin to look for explanatory variables of sex-related differences in mathematical study is in the cognitive area' (Fennema 1979). To date, the only cognitive variable which has been seriously considered to help to explain sex-related differences in mathematical performance is spatial visualisation; much has been written on this factor (for a review, see, for example, Fennema 1979, Bishop 1980, Badger 1982), and it will not be discussed further here, although it would be interesting to explore the connections between spatial visualisation and the more general theory which underlies the arguments advanced in this article.

Gordon Pask and his colleagues have established a strong case for the existence of two distinct learning strategies—serialist and holist; they suggest that learning performance is regulated by the level of uncertainty at which the learner is prepared to operate. Serialists proceed from certainty to certainty, learning, remembering and recapitulating a body of information in small, well-defined and sequentially ordered 'parcels'. They may appreciate topics ahead of those they understand, but they tend not to look far ahead; they are cautious, 'one step at a time' learners who are confident that the necessary knowledge will be gained steadily. Holists, on the other hand, prefer to start in an exploratory way, working first towards an understanding of an overall framework, and then filling in the details; they will tend to speculate about relationships during the learning process and will in general remember and recall bodies of knowledge in terms of 'higher order relations' (Pask 1976a, Holloway 1978). (For a consideration of the relationships between this and other theoretical models of the learning process see de Winter Hebron 1983, Entwistle 1981.)

Of particular importance to this discussion is the fact that in *free* learning situations, where the learner sets her or his own goals, the serialistic and holistic strategies tend to be stable across different tasks for the same learner (Pask 1976b). In order to ensure complete mastery of a complex topic area, teachers must intervene in ways that encourage the learner to adopt a flexible approach; experiments, in which students who were 'fixed' in one mode were provided with evidence illustrating the superiority of another mode, revealed varying degrees of cognitive fixity, and a great deal of 'persuasion' was required to make most students change mode. The degree of cognitive fixity demonstrated is an indication of the versatility of the learner, a versatile learner being one who may adopt either a holistic or a serialistic learning strategy if the subject matter to be learned is changed.

As far as the early content of the mathematics curriculum is concerned, there is no doubt that both serialistic and holistic strategies can be employed with success; in particular cases, however, one approach might prove to be easier or more efficient than the other. In later stages of a child's mathematical experience, when concept areas become more complex, a versatile approach becomes much more important. In problem solving, for example, it may sometimes be the case that reproductive thinking leads to a solution, by means of the existence of stimulus equivalence in the problem situation and in a previously mastered situation; other cases, however, will require productive thinking, whereby past experience is repatterned and restructured to fit the novel situation (Birch and Rabinowitz 1968); a combination of both serialistic and holistic skills is therefore desirable.

It may be significant that Krutetskii, working with a group of mathematically able children, found that, with very few exceptions, these children were demonstrating a versatility of approach (Krutetskii 1976), although it must be noted that he was employing a distinction other than that of holist/serialist. Study of childrens' learning strategies in other areas of the curriculum makes it clear that such versatility can not be taken for granted. In a biological context, for example, Margaret Brumby found that 8 per cent of her sample attacked all tasks in a purely holistic style and 42 per cent used a singular analytic (serialistic) style, with 50 per cent showing varying ability to use both styles of perception in different problems (Brumby 1982).

It is the belief of the writer that children who are predisposed to a serialistic approach are less likely to develop into versatile learners within the mathematics classroom than those who are inclined to adopt holistic strategies; this situation is held to be directly attributable to teacher behaviour.

THE EFFECTS OF TEACHER INTERVENTION IN THE PRIMARY SCHOOL

The discovery methods which are encouraged in primary schools should allow all pupils the freedom to develop their ideas using their preferred learning strategies. However, teachers do sometimes impose their own strategies on their pupils. For example, Ginsburg suggests that children often fail to understand the necessity or rationale for the written procedures they are taught for arithmetic computations, but holds that, despite this, such methods are forced upon them, and in school they are required to use them.

An appropriate strategy for success in arithmetic computation employs the adoption of what Ginsburg calls procedures—algorithmic step-by-step activities (Ginsburg 1983). He defines visually moderated sequences, which have the form of an input (usually visual) that cues the retrieval (from memory) of a procedure; execution of the procedure modifies the visual input; the modified visual input cues the retrieval of a new procedure, and the cycle continues until some process (possibly completing the solution) triggers termination. Such a sequence can be illustrated using a question from the test administered in the survey *Mathematics and the Ten-year-old:*

A visual cue	392 ×7 —— ——
triggers the retrieval of a procedure	(Ah, yes! What are 7 (lots of) 2s)
which produces a new (non-visual) cue	(=14)
which triggers retrieval from memory of another procedure	(I must carry the 1)
which produces a new visual input	392 ×7 —— 4 ——
and the process continues . . .	1

The question above was one of those which the primary school teachers rated as being most important, and also one for which the mean score of the girls is higher than that of the boys. Clearly the method of solution is serialistic in nature, and because it has been shown that learners perform best when the methods being taught are consistent with their own preferred approach, we would expect serialists to perform better than holists on such a task.

In contrast, the area of three-dimensional visualisation provides a context in which it seems that more boys than girls are successful: boys have been (statistically!) significantly better at matching two-dimensional nets to three-dimensional solids (Conner and Serbin 1980, Wattanawaha 1979), and at some of the APU questions in this topic area; for example, in 1979 82 per cent of the boys succeeded at building, from a diagram, a model which needed four hidden blocks to make it stand, compared with 63 per cent of the girls (APU 1980). It is argued here that such problems are easier to do if a 'feel' for three-dimensional shapes has been developed by exploration—a holistic strategy; a serialistic approach to the mastery of the relevant skills is possible, but is much more difficult. (Indeed it is an interesting and worthwhile exercise for holists to attempt to devise an appropriate serialistic strategy.)

It has already been shown that the strategies imposed by teachers for written computation are serialistic in style. It is suggested that it is reasonable to assume that, in fact, the majority of input from most primary school teachers will be serialistic in nature.

The statement above may come as a surprise to those who have followed the trend in primary education towards 'progressive' teaching, where the role of the teacher is to provide an environment in which children are free to explore and discover. In such a situation the teacher does not control the strategies adopted by the pupils—holists will adopt holistic strategies and serialists will adopt serialistic strategies. However, there are still many occasions where the teacher *actively* directs the learning process, and it is at these times that their own strategies become important.

The argument that teacher interventions will be predominantly serialistic is supported by evidence that most primary teachers have a low level of mathematical confidence (Walden and Walkerdine 1982); they also tend not to be competent to a high level in mathematics—for example, in 1974 fewer than 60 per cent of entrants to colleges of education had an 'O' level pass in mathematics (Ward 1979). Walden and Walkerdine report that all the teachers on their primary school study expressed concern that their attempts at explanation might only succeed in further confusing the children! It may be that such teachers are using the theories of progressive education as an excuse for non-intervention in areas of the mathematics curriculum where their own understanding is incomplete.

Discussions with primary mathematics consultants, researchers and lecturers involved in training primary teachers have strengthened the writer's belief that it is still the case that the majority of primary teachers provide input in the classroom which is based on their own (rule-based) experience of school mathematics, and is reflective of a serialistic approach.

The effects of these teacher interventions will be different for holistic and serialistic learners; the serialists will become increasingly committed to the view that a step-by-step approach leads to success in mathematics; even those serialists with enough versatility to become more flexible in other curriculum areas, where they are actively encouraged to adopt other strategies, will fail to do the same in mathematics, because the *value* of alternative approaches will not have been demonstrated; the effect on the holists, on the other hand, will be mediatory; by showing the effectiveness of techniques associated with serialistic skills, the teacher provides the impetus for holists to supplement their self-developed strategies to produce the versatility of approach which underlies complete understanding of mathematical topic areas.

IMPLICATIONS AT SECONDARY LEVEL

Another interesting aspect of Pask's theory, already mentioned briefly, concerns the effect of a mismatch between teaching style/presentation of material and learning strategy. If, for example, the teaching style and materials presented are geared to a holistic approach, it will still be possible (though more difficult) for a serialist to complete a learning task in such a way that his or her level of attainment immediately following completion of the task is equal to that of a holist; in respect of long-term retention, however, such mismatching has a genuine effect in reducing efficiency (Pask and Scott 1972). This disparity may help to explain why some pupils who are successful in other examinations, and perform well in the mathematics classroom, do badly in mathematics examinations. It is believed, however, that there are additional factors arising from Pask's theory which may influence this pattern of performance.

The situation in the primary school has already been discussed. In the secondary mathematics classroom, if the hypothesis under consideration is correct, the pupils will be either versatile or serialistic learners. It could be argued that even at this stage teacher exposition tends to be serialistic in style, and that serialists are therefore not disadvantaged, but such an argument fails to take account of the fact that pupils are expected to do more than simply reproduce items of knowledge as they have been taught. They must, for example, also be able to restructure bodies of knowledge in ways appropriate to different problems — a difficult task for serialists because of their inclination to learn sequentially, without necessarily forming an overall picture of the relationships involved. They will be able to cope with questions set at the time of learning, especially if they can follow the pattern of worked examples, but, if they are tested at a later date, they are unlikely to do well at problems involving restructuring.

A slightly different problem may affect serialists who are following individualised learning schemes like SMILE: Nigel Langdon, addressing students at the Institute of Education, London, likened learning mathematics using SMILE to learning about London using the underground system: you come up at different places, and you explore and learn about small areas; after you have been doing this for some time you begin to realise how different areas relate to each other. A nice analogy, and an interesting idea, but whilst holists are busy speculating about relationships, and discovering the connections between initially disjoint areas of mathematics, it may not even occur to serialists to begin to look for such links; they will be directing their efforts towards the attainment of their immediate goals.

There is no suggestion that serialists are less capable than holists of achieving relational understanding of mathematical concepts; they are failing to do so only because they have not 'learnt to learn' in this way. Pask has designed and tested methods of improving versatility which have proved successful (Pask 1975a and b, 1976a and b), and some recent developments in mathematics education should provide opportunities which encourage serialists to take a more global view: LOGO programming, for example, can provide a context for development of a 'top-down' approach to the solution of problems (Papert 1980). Perhaps a good starting point would be to make explicit to all pupils the desirability of forming the sort of relationships previously mentioned.

ARE MORE GIRLS THAN BOYS SERIALISTS?

Having given a justification for the assertion that serialists are disadvantaged when learning mathematics, it remains to provide a rationale to support the conjecture that more girls than boys fall into this category. The primary reason for supposing that this is probable comes from consideration of the data available from reports of studies concerned with gender differences in mathematics.

The pattern of development of mathematical performance of girls relative to boys is consistent with that which would be expected if there were a higher proportion of serialists amonst the girls: we would expect serialists to excel in written computational skills, but to do less well in areas where they had to apply knowledge to novel situations; it is also predictable that serialists would fall further behind as concept areas became more complex, so that understanding of underlying relationships becomes central to success; this will be particularly important at the higher end of the ability range, and it is here that the ratio of boys to girls is high: the first APU Secondary Survey found that, at age 15/16, 61.5 per cent of the top 10 per cent of achievers were boys (APU 1980b).

In a study of assessment of SMILE students, it was discovered that girls were performing very well in their course work, but less well than boys in examinations (Gmiterek 1984); again we can provide an explanation for serialists displaying such a trait.

Many of the comments used by teachers to distinguish 'boys' success' from 'girls' success' (Walden and Walkerdine) could be used more appropriately to differentiate between versatile and serialistic learners.

It is left to the reader to review other research findings, bearing in mind the versatile/serialist distinction.

Pask does not offer an explanation of the origins of a disposition to adopt one or other learning approach. However, it has already been explained that the level of uncertainty at which individuals are happy to work is a distinguishing characteristic between serialists and holists. It is tentatively suggested that boys' and girls' early experience, and the expectations made of them, may result in more boys than girls adopting methods which are dependent on a willingness to take risks, whilst girls are inclined to be more cautious.

CONCLUSION

Concisely, the hypothesis being put forward for consideration is as follows.

Because of the serialistic nature of teacher input in the primary mathematics classroom, pupils who are predisposed to a serialistic approach are less likely to develop into versatile learners of mathematics than are those inclined to adopt holistic strategies; the most successful mathematics students will generally be versatile learners; the existing evidence concerning differences in performance between girls and boys is consistent with there being a greater proportion of girls (than boys) with serialistic tendencies. This may help to explain why there are fewer girls than boys reaching the highest levels of attainment.

The arguments developed in this article are in many ways simplistic, and it is acknowledged that the position of girls in the mathematics classroom cannot be fully understood without taking account of complex sociological and psychological relationships. However, if the above hypothesis were shown to be valid, it should at least be possible to work towards the elimination of this one factor. It may be that, *because* of the complexity of the situation, the solution to the problems faced by girls in mathematics can best be attacked initially using a serialistic approach!

REFERENCES

Assessment of Performance Unit (APU) (1980) *Mathematical Development*, Primary Survey Report No. 1. London: HMSO.

Assessment of Performance Unit (APU) (1980b) *Mathematical Development*, Secondary Survey Report No. 1. London: HMSO.

Assessment of Performance Unit (APU) (1981) *Mathematical Development*, Primary Survey Report No. 2. London: HMSO.

Assessment of Performance Unit (APU) (1982) *Mathematical Development*, Primary Survey Report No. 3. London: HMSO.

Armstrong, J. (1980) *Achievement and Participation of Women in Mathematics: an Overview*. Education Commission of the States. Denver, Colorado: National Assessment of Educational Progress.

Badger, M.E. (1981) 'Why aren't girls better at maths? A review of research', *Educational Research*, **24**(1), 11–23.

Birch, H.G. and Rabinowitz, H.S. (1968) 'The negative effect of previous experience on productive thinking', in Wason, P.C. and Johnson-Laird, P.N. (eds), *Thinking and Reasoning*. Harmondsworth: Penguin.

Bishop, A.J. (1980) 'Spatial abilities and mathematics education—a review', *Educational Studies in Mathematics*, **11**, 257–269.

Brumby, M.N. (1982) 'Consistent differences in cognitive styles shown for qualitative biological problem solving', *British Journal of Educational Psychology*, **52**, 244–257.

Conner, J. and Serbin, L. (1980) 'Mathematics: spatial ability and sex roles', Report to the National Institute of Education.

De Winter Hebron, C. (1983) 'Can we make sense of learning theory', *Higher Education*, **12**, 443–462.
Douglas, J.W.B. (1964) *The Home and School.* London: Panther.
Entwistle, N. (1981) *Styles of Learning and Teaching.* Chichester: Wiley.
Fennema, E. (1979). 'Women and girls in mathematics—equity in mathematical education', *Educational Studies in Mathematics*, **10**(4), 389–401.
Ginsburg, H.P. (ed.) (1983) *The Development of Mathematical Thinking.* London: Academic Press.
Gmiterek, D. (1984) 'Sex differences in relation to different forms of assessment in a mathematics curriculum project.' Unpublished MA dissertation. University of London Institute of Education.
Holloway, C. (1978) *Cognitive Psychology, Block 4, Learning and Problem Solving (Part 1).* Open University Press.
Joffe, L. (1983) 'Is it your attitude that matters?' Paper presented to GAMMA Conference at University of London Union. Summary in GAMMA Newsletter, November.
Krutetskii, V.A. (1976) *The Psychology of Mathematical Abilities in School Children.* Chicago: University of Chicago Press.
Maccoby, E.M. and Jacklin, C.N. (1975) *The Psychology of Sex Differences.* Palo Alto: Oxford: Oxford University Press.
Papert, S. (1980) *Mindstorms.* New York: Basic Books.
Pask, G. (1975a) *Conversation, Cognition and Learning.* Amsterdam: Elsevier.
Pask, G. (1975b) 'Conversational techniques in the study and practice of education', British Journal of Educational Psychology, **46**, 12–21.
Pask, G. (1976a) *The Cybernetics of Human Learning and Performance.* London: Hutchinson.
Pask, G. (1976b) *Conversation Theory.* Amsterdam: Elsevier.
Pask, G. and Scott, B.C.E. (1972a) 'Learning and teaching strategies in a transformation skill', British Journal of Mathematical and Statistical Psychology, **24**, 205–229.
Pask, G. and Scott, B.C.E. (1972b) 'Learning strategies and individual competence', International Journal of Man–Machine Studies, **4**, 217–253.
Walden, R. and Walkerdine, V. (1982) *Girls and Mathematics: The Early Years*, Bedford Way Papers 8, University of London Institute of Education.
Walden, R. and Walkerdine, V. (1983) *'Girls and mathematics: from primary to secondary schooling.* Unpublished report. University of London Institute of Education.
Walden, R. and Walkerdine, V. (1985) *Girls and Mathematics: From Primary to Secondary Schooling*, Bedford Way Papers 24. London. University of London Institute of Education.
Walkerdine, V. (1983) 'She's a good little worker; femininity in the early mathematics classroom'. Paper presented to GAMMA conference, University of London Union. Summary in GAMMA Newsletter, September.
Ward, M. (1979) *Mathematics and the Ten-year-old.* London: Evans and Methuen.
Wattanawaha, N. (1977) 'Spatial ability and sex differences in performance on spatial tasks.' MEd thesis. Monash University.

6

*Mathematics Learning and Socialisation Processes**

GILAH C. LEDER

Many variables influence the environment in which learning takes place. Models formulated to describe students' learning behaviour in general, and the learning of mathematics in particular, typically include cognitive and affective components as well as socialisation factors.

The impact of the social environment on mathematics learning is described by Reisman and Kauffman (1980) as follows:

> In our culture, mathematics is considered to involve superior reasoning ability and is thought to be a powerful tool. Successful performance in mathematics carries with it positive connotations. Being 'good in math' is 'being bright', and being bright in mathematics is associated with control, mastery, quick understanding. Unsuccessful mathematics achievement implies the opposite of these positive connotations. This value system is a *cultural problem* that has a subtle harmful effect on a number of children and adults.
>
> (Reisman and Kauffman 1980, p.36)

Bishop and Nickson (1983) also acknowledged the relationship between socialisation processes and the learning of mathematics when they argue that research in mathematics education 'should be directed away from the individual child as a learner and towards an increased understanding of the effects of the social context of schools on the learning of mathematics' (p.67).

The socialisation processes operating in schools reflect those of the society within which the schools are placed. They can be quantified in

*An earlier version of this chapter was presented at the fifth International Congress on Mathematical Education, Adelaide, in August 1984.

a number of ways: in terms of cultural expectations, parental expectations and beliefs, school and teacher practices, as well as through peer group pressures. Their collective influence has been highlighted in the debate on sex differences in mathematics learning. Of particular interest, for the purpose of this discussion, is their influence on the learning behaviour of able students.

An examination of the issue of sex differences in mathematics learning from a historical perspective is particularly useful. Not only does this approach exemplify the importance and pervasiveness of different factors on the learning of mathematics, but, in the words of Zeldin, 'that ancient chimera, of using history to understand the present . . . (serves) as a means of being made aware of my own prejudices, of developing a certain detachment' (Zeldin 1981, p.541).

A historical survey of women in mathematics usually discusses the life and work of Hypatia, Emilie du Châtelet, Maria Agnesi, Caroline Herschel, Sophie Germain, Ada Lovelace (especially now computer usage has become so widespread), Mary Somerville, Sonya Kovalevsky and Emmy Noether. Details can be found in Mozans (1913), Osen (1974) and Perl (1978). Typically, the interest shown and encouragement given by at least one close and important male are cited as essential ingredients for the realisation of the mathematical potential of these female mathematicians.

The fathers of Hypatia, Agnesi and Noether were mathematicians who fostered their daughters' interest in mathematics. The Marquise du Châtelet was encouraged in her mathematical pursuits 'by a family friend, M. de Mezières, who recognized her genius' (Osen 1974, p.53) and herself had the financial means to buy high quality mathematical tuition. Much of the work for which Caroline Herschel is remembered was begun when she worked as an astronomical assistant to her brother William who appreciated her help sufficiently to write to the Queen of England in 1787, to ask that his sister be given an allowance of fifty or sixty pounds a year 'by way of encouraging a female astronomer . . . She does it indeed so much better to my liking than any other person I could have, that I should be very sorry ever to lose her from the office' (Turner 1977, p.126). Initial parental opposition to her mathematical studies did not deter Sophie Germain, whose mathematical talents were recognised and encouraged by the well known mathematician Gauss. The close friendship and working relationship between Ada Lovelace (the daughter of Lord Byron) and the English mathematician and inventor, Charles Babbage, have been well documented (Moore 1977). In the case of Mary Somerville freedom to pursue her mathematical studies followed early widowhood that brought financial independence, as well as a supportive second husband whose work-related travels brought her in contact with mathematicians on the Continent. It is also worth noting that her efforts were consistently encouraged by the editor of a mathematical journal (*The Mathematical Repository*). A favourite uncle initially encouraged Sonya Kovalevsky in her mathematical studies. Later, after the university senate refused her admittance to his lectures, the mathematician Weierstrass agreed to help and took her on as his private pupil.

Biographical descriptions of the lives of these female mathematicians show another common theme. Because of prevailing conventions, mathematical studies had to take second place to more stereotyped or sex-role appropriate activities. Emilie du Châtelet is described as an active participant in 'the social life of the court, especially the gambling and the amorous adventures' (Osen 1974, p.54). Her prolonged affair with Voltaire provided the additional benefit of worthwhile intellectual stimulation, however. Maria Agnesi renounced the world of mathematics after her father's death and dedicated the last forty years of her life to charitable projects and in service to the poor, while Caroline Herschel's scientific endeavours were subservient to her brother's research. In a letter to a colleague Sonya Kovalevsky confessed her dual loyalties to mathematics and writing. 'All my life, I have been unable to decide for which I had the greater inclination, mathematics or literature. It is very possible that I should have accomplished more in either of these lines, if I had devoted myself exclusively to it; nevertheless, I cannot give up either of them completely' (Osen 1974, p.137). Mary Somerville's life has attracted the attention of a number of writers. As well as applauding her scientific exploits, they pointed out that 'she was never so narrowly dedicated to them that she disregarded activities in the social, aesthetic, and intellectual world' (Basalla 1963, p. 532).

The overriding influence of cultural pressures on the learning environment can be inferred from a number of other examples. Descriptions of educational programmes for English women in the eighteenth century indicate that as well as singing, dancing, painting and needlework, they might be taught French, reading, writing and sufficient arithmetic for them to be able to keep household accounts. Evidence of the efficiency with which they carried out the latter task abound when historical records of stately homes in England are examined. Within the constraints imposed by society, many women showed proficiency in quantitative tasks. Further evidence is provided by the contents of a little known English periodical, the *Ladies' Diary*. (For a fuller description of this journal see Leder 1981.)

Begun in 1704, the *Ladies' Diary* continued under the same title until 1840, when it combined with the *Gentleman's Diary* until publication ceased in 1871. Throughout its long life, the *Ladies' Diary* contained a mathematical section. Small at first, this section was substantially enlarged in later issues and included mathematical questions specifically addressed to women, as is shown by the following example.

> Dear Ladies fair, I pray declare,
> In Dia's page next year,
> When first it was I 'gan to pass
> My time upon this sphere.
> My age so clear; the first o' the year
> In years, in months, and days
> With ease you'll find, by what's subjoin'd
> Exact the same displays.
>
> $xy + z = 238$
> $xz + y = 158$
> $x + y + z = 39$
>
> Where x = the years, y = the months, and z = the days of my age, the 1st of January, 1795.

This problem can be regarded as a forerunner of the now more common attempts to ensure that mathematics problems found in textbooks and used in teaching appeal to females as well as males.

Some measure of the success that the *Ladies' Diary* had in attracting female readers can be gauged from a comment in the *Diary* of 1718. 'Foreigners, would be amaz'd when I show them no less than 4 or 500 several letters from so many several women, with solutions geometrical, arithmetical (and) algebraical' wrote the editor of that issue. While it is not easy to determine accurately to what extent women contributed to the mathematical section of the *Ladies' Diary* there is considerable circumstantial evidence to indicate that of those who did, a number were the wives or close relatives of men involved in mathematical pursuits. For example, the works of Stephen and Lee (1960) and Taylor (1954, 1966) reveal a number of entries of men who lived at the same time and who had the same surnames as a number of the female contributors to the *Diary*. While such evidence is not conclusive, the relatively large number of 'matching' entries adds weight to the speculation. Patterson (1974) and MacLeod and Moseley (1979) have also argued, independently, that women who were mathematically proficient were likely to have had household exposure to that subject.

In summary, there is a considerable historical evidence to indicate that many females valued exposure to mathematics, readily mastered its contents especially if there was support for this within their family circle, and used the acquired skills most effectively within the bounds set by contemporary society.

Nonetheless, despite the increased educational opportunities currently available to both males and females, sex-stereotyping of certain areas of the curriculum persists. Recently two samples of secondary school students were asked to indicate those adjectives (out of a list of 300) which they considered to be most applicable either to outstanding mathematics students or to outstanding English students (Leder 1985a). Earlier, Williams and Bennett (1975) had used the same list to determine adjectives that were commonly associated with either the male or female role. Many of the adjectives used by Leder's samples to describe both able mathematics and English students (e.g. ambitious, determined, persistent) were those previously defined as being associated with male characteristics. A number of other adjectives most frequently associated with able mathematics students but not with able English students (including logical, rational, quick) also fell into the category. However, approximately half of the adjectives most frequently selected as being descriptive of good English students only (e.g. imaginative, cautious, sensitive) were ones that had been associated with female traits. Thus the stereotypes of good mathematics and English students tapped by the list of adjectives overlapped with male and female stereotypes respectively. Academic success, it seems, particularly in mathematics, rather less frequently in English, still tends to be associated with the male role.

Various models have been proposed to explain sex differences in participation and achievement in mathematics and science. They typically include the variables identified by the historical summary and invariably reflect socialisation factors. A useful overview of a number of these models is given in Leder (1985b), Fennema and Peterson (1985), Eccles (1985), and Maines (1985). Selected details are given below.

Kelly (1981) included cultural expectations and pressures among the process variables in her model describing achievement in science. Fennema and Peterson (1983) argued that students' achievement in mathematics is influenced by their internal motivational beliefs and autonomous learning behaviours as well as by external and societal factors. Variables subsumed under the headings of cultural milieu and socialisers were part of the model proposed by Eccles *et al.* (1983) to describe students' learning and achievement behaviour. The models imply that socialisation experiences affect the mathematics learning behaviour of *all* students. In particular, they stress the importance of these experiences in helping to explain the sex differences in mathematics learning reported in the literature. While it is beyond the scope of this chapter to include a review of this literature, the consistent and recurring finding of sex differences in mathematics learning among able students is particularly worth noting.

Yet another model that adds to the understanding of the influence of socialisation pressures on learning and achievement drives is one proposed within the framework of the expectancy value theory of achievement motivation. In this framework, an individual's achievement behaviour is considered to be a function of that individual's need for achievement, the expectation that the desired goal can be attained, and the value of that goal. A series of carefully designed and executed experiments largely confirmed the above hypothesis provided that the experiments were conducted with male subjects. Contradictory findings, often inconsistent with those obtained with male subjects, emerged from studies with females. Variables that typically aroused achievement motivation in males frequently failed to do so for females. For example, cues that stressed leadership and intelligence qualities aroused optimal achievement efforts in males but did not necessarily have the same effect on females. The latter frequently responded more positively to situations that concerned approval and affection from others. Since achievement motivation was 'assumed to be a multiplicative function of the strength of the motive, the expectancy (subjective probability) that the act will have as a consequence of the attainment of that incentive, and the value of that incentive' (Atkinson and Feather 1966, p.13), presumably sex differences in expectations of reaching the incentive and/or valuation of that incentive helped to account for the differences in achievement arousal.

In an attempt to explain and understand the sex-linked conflicting findings of research on achievement motivation, Horner (1968) postulated the motive to avoid success or fear of success construct. She argued that, since in our culture achievement and the attainment of success in certain areas are considered to be more congruent with the male than the female role, for females the attainment of success may have negative consequences such as unpopularity, guilt, abuse, or doubt about their femininity. Thus fear about the negative consequences that might follow the attainment of success might detract from the value of the goal and might lead to decrements in performance. While fear of success was postulated to be more prevalent in females than in males, it was not expected to be equally important for all women. Fear of success should be more characteristic of high ability, high-achievement-oriented females who aspired to and were capable of achieving success, than of low ability, low-achievement-oriented females who neither desired nor were capable of attaining success, and should be aroused particularly when the tasks involved were generally 'considered masculine such as tasks of mathematical, logical, spatial, etc., ability' (Horner 1968, p.24). An unwillingness to pay the price extracted from those who conspiciously contravene cultural norms may help to explain the lower performance of post-primary-school girls, compared with boys, in mathematics, as well as the consistency of the findings that boys are over-represented among the top mathematics performers.

Typically, assessment of fear of success is through content analysis, according to detailed procedures laid down (Horner *et al.* 1973), of short stories written in response to certain cues administered in a group testing situation. 'Anne/John came top of her/his mathematics class last term', is an example of such a cue. The written stories are assessed for fear of success imagery such as negative consequences because of the success, anticipation of negative consequences because of the success, negative affect because of success, planned activity away from present or future success, direct expression of conflict about success, denial of the situation described by the cue, or bizarre, inappropriate, unrealistic or non-adaptive responses to the situation described by the cue.

As already indicated, many factors are thought to contribute to the environmental climate and socialisation experiences against which individuals have to measure their own aspirations. Instead of enumerating them and identifying the effect of single elements, in this chapter the collective impact of such socialisation experiences are summarised and quantified through a content analysis of relevant articles in the (print) media.

There is much evidence that the media help to shape ideas and attitudes, as well as reflecting and reinforcing popular beliefs (Eysenck and Nias 1978, Roberts and Tyler 1977, Vail 1980). The high penetration of the print media has been well documented. According to American estimates (Guthrie 1979) 90 per cent of the general population read books, magazines, or newspapers. A recent

Australian survey confirmed the high level of penetration of the print media (McNair-Anderson 1984). For example 68 per cent of the people surveyed claimed that they used their local paper to keep informed about local happenings and events. The corresponding figures for radio and television were 16 per cent for each.

Since the print media reflect and reinforce popular stereotypes and since this general climate is reproduced in the education system, it is instructive to examine how various individuals are portrayed in the media. Because capable and achievement-oriented school students are likely to look at outstanding individuals as potential role models, it was decided to concentrate on articles that featured such individuals. It is important to quantify society's perceptions and expectations of them so that student's own life goals can be matched against them. The articles collected were used to obtain the following information:

(a) the range occupations of the individuals portrayed; and
(b) whether or not the articles contained any recurring themes.

METHOD

A press-cutting agency was used to survey the vast array of newspapers, magazines, and periodicals published throughout Australia. The agency used covers all the leading metropolitan daily newspapers and weeklies, as well as many suburban and country papers from all Australian states. Altogether the agency screens over 200 Australian publications. Because of the widespread practice of syndication of features from newspapers published in England and the United States of America the coverage achieved went well beyond Australia.

Articles that featured individuals singled out for their achievement or special accomplishment in any field other than sport were identified. Sport was excluded because of the large numbers of individuals discussed in newspapers' sport sections. The possibility that outstanding sport achievement, particularly of males, would have swamped all other fields of endeavour would thus not arise. One hundred articles featuring 115 females and 35 articles featuring males are discussed here.

RESULTS

The individuals portrayed can be described in terms of Australian Census occupation classifications, as is shown in Table 6.1.

Table 6.1 *Distribution of individuals by selected occupation classification*

Occupational classification	Percentage of females	Percentage of males
Professional and technical	33	34
Administrative, executive and managerial	40	57
Other	27	9

Using these broad and general categories, few sex-related differences appear. It is worth noting that 7 per cent of the females in the 'other' category seemed to have been selected primarily because of their partner's achievements. This did not apply to any of the males.

Quite a few of the individuals selected were in occupations that required some mathematical prerequisites, e.g. pilot, medical researcher, agriculturalist, accountant. During the period monitored, a mathematician *per se* was featured in only one article, part of which is reproduced below.

A TALENT FOR BEING BRAINY

> Nina Morishige was a 'genius' at four, a concert pianist at 16, a mathematician, a champion golfer and the youngest woman to win a Rhodes scholarship to Oxford, England.
> She is brilliant, and knows it. 'Some people are born with beauty,' she says. 'I'm brainy, it's just a talent I have'.
> Miss Morishige has an immaculately tidy mind. She likes mathematics because it 'gets rid of messiness and makes reality a lot tidier and nicer than it appears'. Pinned to her bedroom wall is a remark of Einstein's: 'Make everything as simple as possible, but not simpler.'
> And she has one or two sharp things to say about infant prodigies: 'Some are very elitist. To fit in socially, you mustn't rub people's noses in the fact of your own intellectual superiority.'
>
> (Campbell 1982)

The data can be supplemented with a recently published list of Australia's top 50 thinkers (Milliken *et al.* 1984). Of those selected, 22 had made their mark in science, 28 in the humanities. Six females were included in the list: one in science (a botanist) and five in the humanities.

ANALYSIS OF RESULTS

Content analysis of the articles revealed three common themes.

(a) 'Females need to work harder than males to achieve the goal described';

(b) 'females have to balance success and interpersonal relationships';

(c) 'success somehow happened. It wasn't expected or sought after'.

Some relevant excerpts are reproduced below.

Theme (a): Women need to work harder

> Women who work outside the home have a more difficult time than men . . .
> I could have moved faster and further if I'd been more single minded.
> Self-inflicted pressures to perform well because you're a woman.
> She faces a continuous battle to prove herself. Sees a lot of depreciation of women's abilities.

Theme (b): Success and interpersonal relations

> In common with rest of her contemporaries, she believes her primary role in life is to be a wife and mother.
> Medicine is her main interest. She is happy to lead a single life.
> She does not stay married because she's ambitious and that's scary for men.

Theme (c): Success somehow happened

> She never recognized leadership qualities in herself. Even now she sounds a little surprised that she rose to the top.
> I didn't feel as if I've ever been ambitious . . . everything I've ever done seems to have just happened to me.
> (Source: selected newspaper articles)

The relation between the fear of success imagery and the themes traced in the newspaper articles is illustrated in Table 6.2.

Table 6.2 *Relation between the fear of success themes and the individual components*

Fear of success components	Theme		
	1	2	3
Negative consequences because of success	/		
Anticipation of negative consequences	/		
Negative affect		/	
Planned activity away from present or future success		/	
Direct expression of conflict about success		/	
Denial of the situation			/
Bizarre, inappropriate or non-adaptive responses			/

Approximately one-third of the articles featuring females contained fear of success imagery, compared with only two of those depicting males. Both of these referred to the effect of success on interpersonal relationships.

A chi-squared test indicated that the difference in the frequency of fear of success imagery in the articles focusing on females and males respectively was statistically significant (χ_1^2, = 10.04, P <0.001).

DISCUSSION

The differences between the contemporary portrayal of successful males and females in the media are noteworthy. Society seems to attribute different sources of satisfaction and conflict to them. The portrayals are consistent with and support Horner's fear of success construct. The preoccupation with the negative rather than the positive aspects of success in articles depicting successful females is unfortunate, and may help to explain not only why certain goals have different values for males and females but also why fewer girls than boys concentrate on the still male perceived areas of mathematics and science. Students reading about successful males and females in the print media receive clear messages about the concomitants of success. Educators need to be aware of and sensitive to these images if they are to help all students achieve their potential. Interventions aimed at encouraging able girls to continue with appropriate academic studies need to take account of the way successful females are popularly described if they are to address appropriate and relevant issues. The social context in which mathematics learning takes place cannot be ignored.

REFERENCES

Atkinson, J.W., and Feather, N.T. (Eds) (1966) *A Theory of Achievement Motivation*. New York: Wiley.

Basalla, G. (1963) 'Mary Somerville'. *New Scientist*, **329**, 532.

Bishop, A.J., and Nickson, M. (1983) *A Review of Research in Mathematical Education. Part B: Research on the Social Context of Mathematics Education*. Windsor, Berks.: NFER & Nelson.

Campbell, J. (1982) *A Talent for Being Brainy*. Launceston: Examiner Express. 13/3/82.

Eccles, J. (1985) Model of students' mathematics enrolment decisions. In Fennema, E. (ed.) 'Explaining sex-related differences in mathematics: theoretical models'. *Educational Studies in Mathematics*, **16**, 303–320.

Eccles, J., Adler, T.F., Futterman, R., Geoff, S.B., Kaczala, C.M., Meece, J.L., and Midgley, C. (1983) 'Expectancies, values and academic behaviors'. In Spence, J.T. (ed.) *Achievement and Achievement Motivation*. San Francisco: W.H. Freeman.

Eysenck, H.J., and Nias, D.K. (1978) *Sex, Violence and the Media.* New York: Harper & Rowe.

Fennema, E., and Peterson, P.L. (1983) *Autonomous learning behavior: A possible explanation of sex-related differences in mathematics.* Paper presented at the annual meeting of the American Educational Research Association, Montreal.

Fennema, E., and Peterson, P.L. (1985) 'Autonomous learning behavior: a possible explanation of sex-related differences in mathematics.' In Fennema, E. (ed.) 'Explaining sex-related differences in mathematics: theoretical models'. *Educational Studies in Mathematics*, **16**, 303–320.

Guthrie, J.T. (1979) 'Why people (say they) read: Reading activities of adults.' *The Reading Teacher*, **32**, 752–755.

Horner, M.S. (1968) *Sex differences in achievement motivation and performance in competitive and non-competitive situations.* Unpublished doctoral dissertation, University of Michigan.

Horner, M.S,, Tresemer, D.W., Berens, A.E. and Watson, R.I., Jr. (1973) *Scoring Manual for an Empirically Derived Scoring System for Motive to Avoid Success.* Unpublished manuscript. Harvard University.

Kelly, A. (ed.) (1981) *The Missing Half.* Manchester: Manchester University Press.

Leder, G. (1981) The *Ladies' Diary. The Australian Mathematics Teacher*, **37**, 3–5.

Leder, G.C. (1985a) *Profiles of able mathematics students.* Paper presented at the 6th world conference on gifted and talented children, Hamburg.

Leder, G.C. (1985b) 'Sex-related differences in mathematics: An overview.' In Fennema, E. (ed.) Explaining sex-related differences in mathematics: Theoretical models. *Educational Studies in Mathematics*, **16**, 303–320.

McNair-Anderson. (1984) 'A survey conducted for the Victorian and N.S.W. Country Press Association.' *Mansfield Courier*, 10/5/84.

McLeod, R., and Moseley, R. (1979). Fathers and daughters: Reflections on women, science, and Victorian Cambridge. *History of Education*, **8**, 324–333.

Maines, D.R. (1985) Preliminary notes on a theory of informal barriers for women in mathematics. In Fennema, E. (ed.). 'Explaining sex-related differences in mathematics: theoretical models'. *Educational Studies in Mathematics*, **16**, 303–320.

Milliken, R., Smith, D., Johnson, S., and Rodell, S. (1984) 'Australia's top 50 thinkers.' *National Times*, 26/5/84.

Moore, D.L. (1977) *Ada, Countess, of Lovelace.* London: John Murray.

Mozans, H.J. (1913) *Women in science.* New York: Appleton & Co.

Osen, L.M. (1974) *Women in mathematics.* Cambridge, Mass.: MJT Press.

Patterson, E.C. (1974) 'The Case of Mary Somerville: An Aspect of Nineteenth Century Science'. Proceedings of the American Philosophical Society, **118**, pp. 269–275.

Perl, T. (1978) *Math Equals: Biographies of Women Mathematicians and Related Activities.* Menlo Park: Addison-Wesley.

Reisman, F.K., and Kauffman, S.H. (1980) *Teaching Mathematics to Children with Special Needs.* Columbus Ohio.: C.E. Merrill Publishing Co.

Roberts, D.J., and Tyler, I.K. (1977) 'Bridging the gap between schools and

communities.' *Journal of Research and Development Education*, **10**, 15–25.
Stephen, L., and Lee, S. (eds) (1960) *The Dictionary of National Biography*. London: Oxford University Press.
Taylor, E.G.R. (1954) *Mathematical Practitioners of Tudor and Stuart England*. Cambridge: University Press.
Taylor, E.G.R. (1966) *The Mathematical Practitioners of Hanoverian England*. Cambridge: University press.
Turner, A.J. (1977) *Science and Music in Eighteenth Century Bath*. Bath: Mendip Press.
Vail, P.L. (1980) *The World of the Gifted Child*. London: Penguin.
Williams, J.E., and Bennett, S.M. (1975) 'The definition of sex stereotypes in the Adjective Check List.' *Sex Roles*, **1**, 327–337.
Zeldin, T. (1981) 'After Braudel.' The *Listener*, November.

7

*Gender Roles at Home and School**

ALISON KELLY *with* Juliet Alexander, Umar Azam, Carol Bretherton, Gillian Burgess, Alice Dorney, Julie Gold, Caroline Leahy, Anne Sharpley and Lin Spandley

The links between children's home experiences and their school behaviour and performance have long been of interest to sociologists of education. They have provided a theoretical battlefield for the nature/nurture debate and the deficit/difference argument. Latterly the discussion has continued in terms of cultural capital and cultural reproduction. Interest in gender divisions in education is more recent, but now constitutes a flourishing field of study. But these two research traditions have had little impact on each other. Modern collections of empirical work, such as Craft, Raynor & Cohen (1980), still frequently ignore gender roles and women's employment. Feminist theorists such as David (1980) and MacDonald (1980) have explored the 'family-education couple' and examined the implications of reproduction theory for understanding women's education. Yet their work is necessarily tentative because of the shortage of relevent empirical studies.

There is now a considerable body of work on socialisation in early childhood (see Maccoby & Jacklin 1975, for a review) and a number of studies of gender roles in school (e.g. Clarricoates 1980, Delamont 1981, Llewellyn 1980, McRobbie 1978, Sharpe 1976, Wolpe 1977). Yet the relations between gender socialisation at home and at school remain unexplored. As Arnot (1981) points out, in these studies an 'unexamined assumption is that the family and the school transmit

*This chapter first appeared in the *British Journal of Sociology of Education*, **3**(3), 1982. Reproduced with permission.

the *same* definition of gender and that no conflict occurs between these two social institutions'. With a few exceptions, gender socialisation is seen as a continuous process at home and school, with little variation by class or race, and few contradictions for the children involved.

But this is not necessarily the view of the schools. In discussions with secondary school staff about sex typing it is soon obvious that many teachers feel constrained by what they perceive to be the opinions of their pupils' parents. Borley (1982) recorded the following comments from head teachers:

> Parents have enormous influence, but it is usually a sex-stereotyped influence because parents have more weight than teachers. Parents need to be educated because they have a very negative attitude.
> Parents' influence on the choice between history and geography is very little, but when it comes to boys' and girls' subjects their indirect influence is enormous due to their own backgrounds. The parents' attitude is the most old fashioned; like schools were twenty-five years ago. They are out of touch.

Yet few schools have made any systematic effort to discover the parents' views, far less to modify them. Their opinions are often based on the vociferous objections of one or two parents (usually fathers of boys) to arrangements whereby all pupils do some technical craft and some domestic craft in lower school.

The question of how widespread these objections are, and how severe a constraint parental opinion really places on the ability of schools to influence children's gender roles is explored below. The paper also considers the contradictory messages about gender roles which children may be receiving from their homes, and the different expectations which working-class and middle-class parents may hold for their daughters and sons.

DATA

The study was carried out at one of the schools involved in the Girls Into Science and Technology (GIST) project. GIST is an action research project which is trying to break down traditional sex roles in school and encourage more girls to choose scientific and technological subjects when they become optional (see Smail, Whyte & Kelly 1982 for further details). First-year pupils in the ten schools involved have completed a number of questionnaires on their attitudes to science, to sex roles and to home and school. These will be compared with the children's subject choices in their third year and their reactions to GIST interventions. Parental opinions and behaviours are often assumed to have an important influence on children's choices; so we were anxious to obtain some information from the parents of children in the GIST project to supplement the information from the children themselves.

The Head and Deputy Head at Tall Trees School (both women) are strongly committed to the aims of the GIST project. They agreed that parental attitudes were an important influence on the children in their school and that it would be interesting to investigate these. They provided us with a list of the names and addresses of all the first-year pupils at the school. They also made helpful comments on, and approved the final versions of, the questionnaire and interview schedule we used. Apart from this the study was conducted through Manchester University by a group of students on a survey research course. Respondents were assured that no information on their individual replies would be available to the school.

It was decided not to tell the parents that we were specifically interested in sex roles, since we feared that this might lead them to give what they thought were 'correct' or 'expected' answers. Nor did we ask for their general opinion about what girls should do and what boys should do, since we were primarily interested in their hopes and expectations for their own child. Instead we described the study as one which explored the links between home and school and allowed the parents to express their views on what the school should be doing. Parents were asked to answer the questions with reference to the child who had just started at Tall Trees School; responses were then analysed according to the sex of that child.

There were 152 pupils in the 1980 intake to Tall Trees School. The parents of 36 of these were selected for interview and the remaining 116 were sent postal questionnaires. Eighty per cent of the questionnaires (92 in all) were returned after one reminder and 24 interviews were completed. The questionnaires and interviews covered similar ground, although more detailed probing of parental attitudes was of course possible in the interview. Because of the larger sample size and the more quantifiable nature of the data most of the results reported here have been taken from the questionnaire study, with supplementary material (mainly quotations) from the interviews.

Tall Trees School is a well-established 11–16 co-educational comprehensive school in an inner suburb of an old industrial city. It has a mixed catchment area of owner-occupied semis and council housing, including some flats. Approximately 1 per cent of the children are non-white, 12 per cent are from single parent families, 80 per cent of their mothers have a job, and 8 per cent of their fathers are unemployed. The children were divided into middle and working class on the basis of their parents' occupations. The occupations were grouped according to the Registrar General's classification into: (1) professional and managerial, (2) intermediate, (3a) white collar, (3b) skilled manual, (4) semi-skilled manual, (5) unskilled manual. The family was considered middle class if either the father had a non-manual occupation (groups 1, 2 or 3a) or the mother had a professional, managerial or intermediate occupation (groups 1 or 2). This definition was adopted in preference to the traditional reliance on father's occupation, so as to give due weight to status derived from mother's employment, while recognising that the 'paper production line' staffed by female typists and shop assistants (group 3a) has lower prestige and reward than traditional male white collar jobs. Using this definition 57 per cent of the returned questionnaires came from working-class families and 43 per cent from middle-class families. (Using father's occupation alone, 59 per cent were working class and 41 per cent middle class.) These proportions were very close to those obtained by applying the same definitions to parental occupations as given by the children in questionnaires completed at the school. This suggests both that working-class and middle-class families were equally likely to return the questionnaire, and that the children were reasonably accurate in reporting their parents' occupations. The only major discrepancy was that 80 per cent of the children said their mothers were employed, but only 64 per cent of the returned questionnaires came from families where the mother was employed. Perhaps working mothers and their husbands are too busy to fill in questionnaires!

Although the pupils at Tall Trees School and their parents are clearly not a random sample, there is no reason to suppose that they are very different from pupils and parents in similar neighbourhoods elsewhere. When significance tests are quoted in this report they can be taken to indicate the possibility of generalising these results to other similar school populations. But such generalisations must of course be treated with caution because of the non-random nature of the samples.

EDUCATIONAL ASPIRATIONS

Most of the parents who responded to the questionnaire attached great importance to their children's education and had high educational aspirations for them. The level of interest was manifested in the high response rate (80 per cent) as well as in the actual answers. Not one parent wished their child to leave school as soon as possible without any educational qualifications. Over 50 per cent wanted them to go on to full-time college or university (see Table 7.1). Parents attached just as much importance to girls' education as to boys'—indeed more girls' parents (84 per cent) than boys' parents (73 per cent) returned the questionnaire and parents were more likely to want their daughters to go on to college than their sons.

Table 7.1 *The percentage of girls' and boys' parents who want their children to leave education at various stages*

	Middle class		Working class		
	% Girls	% Boys	% Girls	% Boys	% All
After CSE/'O' level	14	15	24	35	21
After 'A' level	9	38	28	25	25
Go to college/university	77	46	48	40	54
N	22	13	25	20	89

The familiar pattern of working-class parents attaching lesser importance to education than middle-class parents emerged clearly in this study. Working-class parents were significantly more likely to agree that 'the things children learn in school are not much use after they've left' and to disagree that 'it is important to get a good education while you are young' and that 'I wish I'd worked harder when I was at school' than middle-class parents. They also had lower aspirations for their children to get qualifications (see Table 7.1). However these differences were only slight and it must be stressed that the overall endorsement of the value of education was very high.

Both middle-class and working-class parents had slightly higher educational aspirations for their daughters than their sons. This sex differentiation was evident at 'A' level for the middle classes and at CSE/'O' level for the working classes (see Table 7.1). It may well reflect the different labour markets for girls and boys. Parents seem to realise that boys can earn good money and have good prospects by leaving school and taking on-the-job training (an apprenticeship or day release) but that these opportunities are traditionally closed to girls, who need academic qualifications if they are to advance. However, there is no evidence here to support the traditional view that parents push their sons more than their daughters at school.

When asked to respond directly to the statement that 'it is more important for boys to get a good education than for girls', two-thirds of the parents disagreed strongly. Parents of boys were more likely to agree than parents of girls, especially if there were no girls in the family. In fact responses to this statement were highly polarised with 20 per cent of parents agreeing strongly that boys' education was more important, and very few being undecided. The reasons given for agreeing were predictably that

> men are the breadwinners, women marry and have children.

But even some of the reasons for disagreeing had the same flavour

> The boy will eventually be the breadwinner—But a girl's education is not wasted—a good education is just as important for girls, mainly to help them set up a good home.
> Both need a good education, so that if anything happened to the husband the wife can go out to work.

Many others indicated that they thought conditions were changing and equality was relatively recent.

> I used to, but not now because there are so many new career opportunities for girls.
> Not now, women can go a lot further now than at one time.
> Girls have and need careers these days.

Thus despite their high ambitions for their daughters, many parents are still not sure that as a general principle girls need education on the same terms as boys.

High parental interest in their own children's education is also evident in the replies to questions about attendance at parents meetings and help with homework. Only 18 per cent of parents said they never attended parents meetings and most of these gave reasons such as working nights, being a single parent or having illness in the family. The level of attendance was markedly higher in the middle class than in the working class, particularly for fathers. Only 16 per cent of children never got help with maths at home and only 26 per cent were never helped with English. Help was generally available whenever the child asked for it, and in many cases when the child was not helped this was because he or she was seen as being too good at the subject to need any help. The way in which help with homework was allocated is shown in Table 7.2. Fathers helped more than mothers with maths, whereas the reverse was true for English. This suggests that parents are displaying sex-stereotyped competencies. Moreover the boys seem to be more aware of this than the girls. Boys show a distinct tendency to get help from their fathers in maths and help from their mothers in English, while the girls' help is more evenly distributed.

Table 7.2 *The percentage of girls and boys receiving help with maths and English from their parents*

	Maths			English		
	% Girls	% Boys	% All	% Girls	% Boys	% All
Help from mother	31	28	29	52	63	57
Help from father	53	61	56	52	37	46
No help	13	21	16	29	21	26
N	53	38	91	52	38	90

Note: The percentages total more than 100 because respondents were permitted to name more than one source of help for each subject. A small number named people other than parents, usually siblings (not shown).

At the beginning of the questionnaire parents were asked what they thought the school should be teaching. The first question, an open-ended enquiry as to the most important subjects for the child to learn, produced an almost unanimous response: maths and English. The answer to the next question, whether the scheme operating at Tall Trees School where all children study both technical crafts and domestic crafts in the first two years, was desirable, was also nearly unanimous. Ninety-three per cent of parents said yes. This result must be encouraging for schools which have already introduced such craft circuses, and reassuring for those which are thinking of doing so but fear parental opposition. When asked why they liked this arrangement most parents gave social reasons

> it will help them to be independent;
> a bit of everything helps if John ever has to live on his own,
> it helps them to look after themselves if necessary;
> it gives a knowledgeable interest he or she can share with brothers or sisters;
> it's a good idea for each sex to see what the other sex normally does in life.

Only 14 per cent mentioned increased job opportunities, and this was more common for boys than for girls. This very limited recognition of the occupational relevance of technical craft for girls is disappointing, since one of the main motivations of the GIST project is the possibility of opening up some traditionally male occupations to girls. As with the question on the importance of education it seems that a superficially egalitarian response may conceal the different connotations of craft subjects for girls and boys.

The importance of specific subjects was explored more systematically in the next question. Parents were asked to rate, on a scale from five (extremely important) to one (a waste of time) how important they thought it was for their child to continue various subjects when

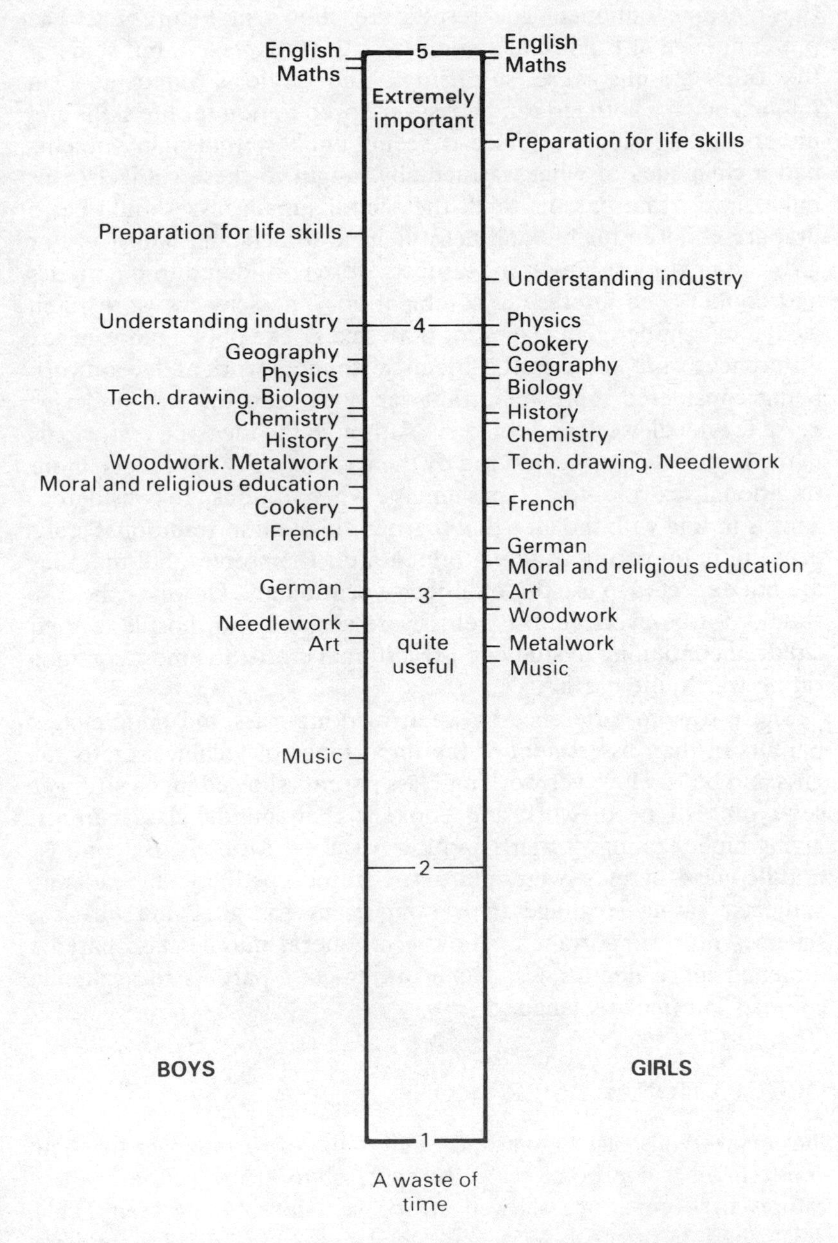

Figure 7.1 Mean parental ratings of how important school subjects are for their children to study.

they became optional. The results are shown in Figure 7.1. The pre-eminence of English and maths was to be expected, but some of the other results were surprising. The subjects rated next in importance for both girls and boys were preparation for life skills and understanding industry. Since it seems unlikely that many parents had a clear idea of what was actually taught in these subjects, this might imply that parents think the school's main task should be to prepare children for life and help them to understand industry. For girls the next most important subjects were considered to be physics and cookery, an interesting combination. The sciences were rated well above modern languages for both sexes. The only significant sex differences were in the craft subjects, with metalwork and woodwork being considered more important for boys than for girls and vice versa for needlework and cookery. Although the idea of a craft circus in the first two years is endorsed by the vast majority of parents, more traditional sex-role stereotypes emerge when options are considered. This is in line with the idea that a grounding in non-traditional crafts is useful in preparing children to look after themselves, but that they are not expected to use these subjects vocationally. However the case should not be overstated. Parents were certainly not hostile to their children continuing to study non-traditional crafts; the most common rating was 'quite useful'.

There was no difference between working-class and middle-class parents in their assessment of the importance of technical crafts for girls and boys. However working-class parents showed much stronger sex-typing of needlework and cookery than middle-class parents, giving higher ratings for girls and lower ratings for boys. By contrast middle-class parents were more sex-stereotyped about academic subjects, rating languages more important for girls and physical sciences more important for boys. In general middle-class parents attached more importance than working-class parents to academic subjects, particularly languages.

OCCUPATIONAL ASPIRATIONS

Parents were also asked to rate the suitability of various jobs for their children on a five point scale. In contrast to school subjects their ratings of occupations showed strong sex-stereotyping (see Table 7.3). The jobs of nurse, secretary, social worker and hairdresser were considered much more suitable for girls than for boys; the jobs of engineer, electrician and draughtsman were considered much more suitable for boys than for girls.

Table 7.3 *Mean parental ratings of the suitability of various occupations for their children*

	Girls		Boys
Nurse	3.1	***	1.5
Doctor	2.7		2.3
Engineer	2.0	***	3.9
Secretary	3.8	***	1.4
Manager	3.0		2.9
Social worker	3.7	***	2.6
Electrician	1.7	***	3.7
Draughtsman	2.4	***	3.6
Shop assistant	2.2	*	1.6
Factory worker	1.3	**	1.9
Hairdresser	3.0	***	1.4
Teacher	3.8	*	3.1
Computer operator	3.6		3.7
N	51		37

***$p<0.001$; **$p<0.01$; *$p<0.05$.

The only jobs which were not sex-stereotyped were doctor, manager and computer operator—despite the fact that the vast majority of doctors and managers are men. This suggests that female professionals may gain acceptance more easily than female craftspeople—perhaps because they keep their hands clean. Computer operator is anomalous at first sight, but it seems likely that many parents do not distinguish clearly between different grades of work with computers, and think that a computer operator is a professional. Overall the ratings for jobs were quite low, certainly by comparison with the ratings for school subjects. This suggests the parents are seriously considering the aptitudes and interests of their own child and not just responding to the social prestige of a job. Class differences were few, although middle-class parents were more favourably disposed than working-class parents towards their sons becoming managers or teachers.

Parents were also asked the open-ended question 'what sort of job would you like your child to get?' Over half the respondents didn't answer this question, or gave vague replies such as

> whatever she thinks will make her happy;
> I would like him to get a job which he chose himself and which he really liked.

But those who were more specific again had very high aspirations for their children. Seventy-six per cent wanted their daughters to get middle-class jobs, mainly nursing or teaching; 46 per cent wanted middle-class jobs for their sons with another 46 per cent wanting skilled manual jobs. Over 60 per cent of parents did not want their child to do the same job as themselves. When asked to give reasons, the child's happiness and financial security vied with each other for priority:

> one in which he will be happy and financially secure;
> a profession, so that she will have a better standard of living than I;
> one with an apprenticeship because this would give him some security.

However it was noticeable that good prospects and security were seen as particularly important for boys, while interesting work and the child's own preference were more commonly mentioned for girls. This distinction carries with it the implication that boys have to work as providers, but that girls work is less serious.

Parents' ideas about the sex-typing of occupations were explored more thoroughly in the interviews. Only 18 per cent of parents thought that men and women should definitely do different jobs, but 41 per cent gave qualified replies. Most of these indicated that there were some jobs they thought women could not do because of the physical stress involved (e.g. mining, lorry driving). When they were asked whether they would be disappointed or embarrassed if their own child took a job stereotypically associated with the opposite sex (motor mechanic for girls, typist for boys) 87 per cent said they would not. Some were quite enthusiastic:

> no, she's quite interested in that sort of thing;
> no, his elder brother enjoys typing;

while others were merely accepting:

> no, if that is what he wants to do, but it seems a bit odd.

This accords with the tendency noted by Newson *et al.* (1978) for parents to support their children's wishes, even if they consider them rather eccentric.

Only one parent was against equal pay for equal work. But several stressed that the man should be the main breadwinner in the family and suggested that women should only work if it was financially necessary. As might be expected from this sample, with its high proportion of working mothers, most parents approved of mothers taking employment outside the home and men doing a share of the housework. But many specified that mothers should work

> only if they don't neglect the kids.

Again it is obvious that parents' ideas on sexual equality presuppose that girls and boys will have different adult roles, and the equality is relative to these roles.

THE THEORY AND PRACTICE OF EQUALITY

The first part of the postal questionnaire asked parents about their own children. In the second part they were asked to respond to a set of general statements about women and men, girls and boys. These were intended to assess parents' theoretical position on sex roles, for comparison with the practical example set by their behaviour and attitudes towards their own children.

The statements on sex roles were interspersed with others on the value of education and the importance of learning science. Parents were asked to indicate their agreement or disagreement with each one on a five-point scale. The responses were then combined into a scale named Sexist (see Table 7.4). High scores indicate more

Table 7.4 *Items included in the Sexist scale with some item and scale statistics*

	Mean	Standard deviation	Item–scale correlation
1. A women's place is in the home.	1.8	1.4	0.39
2. Boys are usually better than girls at maths	2.2	1.4	0.47
3. Men who do housework are a bit soft	1.4	1.1	0.60
4. Boys make better leaders than girls	2.2	1.5	0.57
5. It is more important for boys to get a good education than for girls	2.1	1.6	0.31
6. Women's lib is all nonsense	2.5	1.5	0.55
7. A man should always be the boss in his own home	2.3	1.6	0.45
8. Girls are usually better than boys at English	2.3	1.4	0.32

Sexist scale: Mean 16.4
Standard deviation 6.9
Possible range 8–40
Actual range 8–34
Cronbach's alpha reliability coefficient 0.76

traditional attitudes about sex appropriate behaviour than low scores. In fact most of the statements on the scale were quite extreme and the general tendency was to disagree with them. Nevertheless some interesting patterns emerged (see Figure 7.2).

Working-class parents scored significantly higher on the Sexist scale than middle-class parents. This is in contrast to the very limited class differences in parents' ideas about what subjects and jobs are appropriate for their daughters and sons. It may represent a real difference in the way parents from different socio-economic groups think about sex roles; or it may merely indicate that middle-class parents are more sensitive about expressing sentiments which could be construed as sexist.

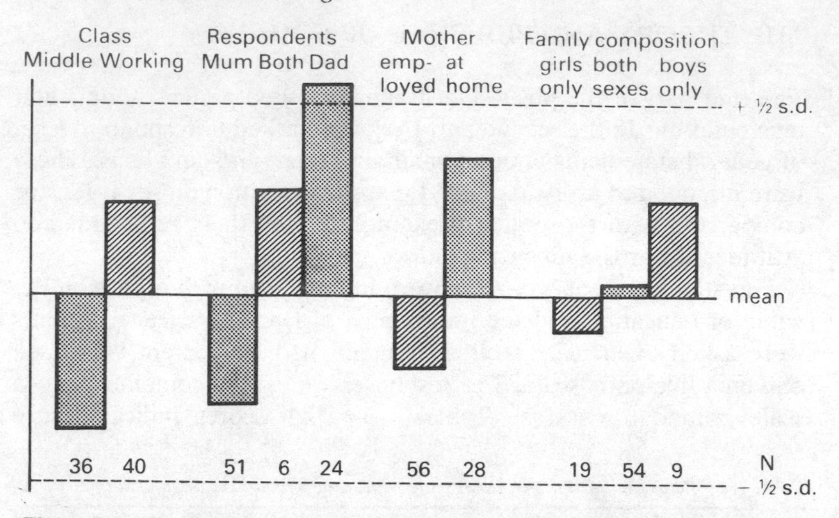

Figure 7.2 The average score on the sexist scale of different categories of parents.
Note: The actual score on the sexist scale has little meaning, so groups are shown in relation to the overall mean and standard deviation of parents' score.

When the postal questionnaires were sent out they were addressed to the 'Parent or Guardian of . . .' with no further instruction as to who was to complete them. In fact the majority were completed by mothers, about a quarter by fathers, a few by both parents together, and three by other people (grandparents and siblings). Figure 7.2 shows that the fathers who replied expressed much more sexist ideas than the mothers who replied, with parents who answered together being intermediate. At first sight this seems to suggest that men are more sexist than women. This may be true, but there is another possible explanation: the fathers who replied may have been more than usually dominant within the family (as evidenced by their taking over a task more usually done by mothers) and their views may not be representative of those of other men. Fathers who replied were slightly more likely to be working class and have daughters.

Respondents from families where the mother was employed scored significantly lower on the Sexist scale than those from families where the mother was a full-time housewife. The pattern was the same whether the questionnaire was answered by the mother or the father. Mothers with younger children were less likely to be employed, but 26 per cent of those with no children younger than 11 were not employed. This suggests that in some families at least the existence of a full-time housewife is a conscious endorsement of a traditional female role. This is a classic chicken-and-egg situation: perhaps they believe in traditional sex roles and so the mother stays at home; or perhaps the mother stays at home for other reasons, such as caring for an elderly relative, and adopts an ideology which justifies her position. Alternatively it may be that the experience of employment and the realisation that the woman is contributing to the family income have a radicalising effect on couples where the mother goes out to work. Maternal employment was not related to class.

These statements were framed in general terms and did not apply specifically to the child in the first year of Tall Trees School. For this reason the results were not analysed by the sex of that child. However parents' ideas may well be influenced by the composition of their family. Parents of girls may be more conscious than parents of boys of the ways in which sexism inhibits their children's opportunities. Although most of the families we studied contained both girls and boys, this idea did receive some support. Parents who had only girls scored lower on the Sexist scale than parents who had only boys (see Figure 7.2).

Scores on the Sexist scale were also analysed by the age of the respondent (not shown). There was little difference between parents in the younger age groups, but the four respondents over 50 scored markedly higher than the others. If this is a true generation gap it may show that attitudes towards sex roles are becoming more liberal.

Parents' sex stereotypes are not just abstract ideas; they can take concrete forms. One of these is the household tasks that children are made to help with. In the interview parents were asked whether they thought girls and boys should do the same household jobs, and 79 per cent said yes. But when the parents answering the questionnaires were given a list of household jobs and asked to mark those with which their child regularly helped, the results were quite strongly differentiated by sex (see Table 7.5). The most striking division was in the percentage of girls and boys doing the washing-up. It is commonly supposed that this job has become the province of 'helping' fathers and husbands. But if the results of this survey are anything to go by, the next generation of boys is not being trained for this role. Only 29 per cent of these 11-year-old boys, compared to 72 per cent of the girls, regularly helped with the washing-up. A significantly higher proportion of girls also helped with cleaning the house, cooking, and tidying their own rooms. And girls were more likely than boys to be found shopping, washing and mending clothes and laying the table. Boys more often than girls cleaned windows, took out the rubbish, washed the car, did the gardening, helped with minor repairs and cleaned shoes.

Table 7.5 The percentage of girls and boys who regularly help with various household tasks

	% Girls		% Boys
Washing up	72	***	29
Cleaning the house	46	*	21
Cleaning windows	2		5
Shopping	76		68
Taking out the rubbish	33		37
Cooking	33	*	10
Washing the car	15		34
Gardening	15		34
Washing clothes	9		0
Mending clothes	9		0
Tidying own room	85	**	55
Laying table	59		42
Minor repairs	4		13
Cleaning shoes	37		50
N	54		38

***$p<0.001$; **$p<0.01$; *$p<0.05$.

Children did not, in general, spend very much time on these household tasks, but girls were more heavily committed than boys. Seventeen per cent of girls and only 3 per cent of boys spent more than three hours a week on household jobs; the corresponding figures for less than one hour per week were 33 per cent of girls and 49 per cent of boys. Children whose mothers were full-time housewives actually spent slightly longer each week on household tasks than those with employed mothers, and boys with mothers at home were much more likely to do the washing up. Perhaps it takes longer to make boys do washing-up than to do it oneself! There were no class differences in the overall time girls and boys spent on household tasks, but middle-class boys were even less likely to do a regular share of the washing-up than working-class boys.

The girls also spent slightly more time than the boys doing homework, although again this was not excessive. Thirty-five per cent of girls and 40 per cent of boys spent less than an hour each evening on homework and only 6 per cent of girls and no boys spent more than two hours each evening on prep. Eighty-one per cent of parents thought that their children did the right amount of homework, although 16 per cent thought they did too little. Not surprisingly these were mainly the parents of children doing less than an hour each evening; only two parents, both with daughters studying for between one and two hours a night, thought their children did too much homework.

Table 7.6 *The percentage of girls' and boys' parents mentioning their child's enjoyment of different types of book*

	% Girls		% Boys
Fiction	83		66
Science fiction	4	*	24
Comics/magazines	6	**	29
Sports/hobbies	2	***	29
Physical science or technology	0	*	13
Biological science	13		10
Other non-fiction	8		24
N	52		38

***$p<0.001$; **$p<0.01$; *$p<0.05$.

Girls and boys were also differentiated in their voluntary activities. Because of the importance of reading for school achievement the questionnaire was quite detailed on this topic. On average girls visited the library more often than boys and spent more time reading for pleasure. But the largest differences were in *what* they read (Table 7.6). This was an open-ended question where parents were asked, without prompting 'What type of book does your child enjoy most?' Fiction was the most commonly mentioned category for both sexes, but science fiction, comics and magazines, sports and hobbies and physical science or technology were all mentioned significantly more often for boys than for girls. The implications of this for schooling are not clear. Although their background reading in physical science and other non-fiction may give boys a useful grounding in some subjects, the relevance of science fiction, comics, magazines, sports and hobbies is less obvious. And girls may benefit simply from the greater quantity of reading that they do.

Parents were also asked the open-ended question 'What else does your child like to do?' Again sex-typed voluntary behaviour was evident (Table 7.7). Very few children of either sex were interested in what might be described as stereotypically female play (e.g. dolls, skipping). Probably this sort of activity is characteristic of a younger age group. However, a large number of boys and very few girls participated in stereotypically male play (e.g. football, fishing, war games). On the other hand a large number of girls and very few boys were mentioned as liking to do stereotypically female jobs (e.g. cooking, knitting); while few children of either sex did stereotypically male jobs (e.g. mending the car) for fun. Other sex differences were minimal; both girls and boys played a lot of sport but other activities were less frequently mentioned.

Table 7.7 *The percentage of girls' and boys' parents mentioning their child's participation in different spare-time activities*

	% Girls		% Boys
Stereotypically female play	9		0
Stereotypically male play	4	***	50
Stereotypically female jobs	31	**	3
Stereotypically male jobs	2		3
Playing sport	56		60
Watching TV	13		26
Intellectual hobbies	30		26
Youth culture	22		18
N	54		38

***$p<0.001$; **$p<0.01$

We did not ask the parents what household jobs they themselves did or what their own spare-time activities were. However we did ask the children about their parents' activities in the questionnaire they completed at school. The results showed that the parents make strongly sex-typed role models (Kelly, Smail and Whyte, 1981). Eighty-four per cent of mothers and 14 per cent of fathers regularly go shopping for food; 87 per cent of mothers and 17 per cent of fathers regularly cook meals; 85 per cent of mothers and 4 per cent of fathers regularly wash clothes; 80 per cent of mothers and 8 per cent of fathers regularly clean the house. Although they say that men should share the housework, especially when their wives are working, very few of these families seem to be practising what they preach.

DISCUSSION

The picture that emerges from this study is one of ambivalence and confusion. Most parents are committed to some form of equality between the sexes. Parents wanted their daughters and sons to study broadly the same subjects at school and they generally rejected the idea that boys' education was more important than girls'. They were consistently in favour of equal pay for men and women and most parents approved of women and men entering the same occupations and sharing the housework. They also thought that mothers should have the chance to work if they wanted to.

However this formal commitment to equality coexisted with other attitudes which could make true equality impossible. Most parents accepted without question that men and women have fundamentally different roles in the family. Children are mainly the woman's responsibility and they are her main responsibility. Women have the right to work, but only if they can do so without detriment to their maternal and domestic roles. These parents do not see equality of the sexes as necessitating any great change in the way society is organised. Equality is a little something extra, which can be added on to existing social arrangements.

Parents' expectations for their own children reveal another dimension of their attitudes. This is the taken-for-granted, lived reality of sex roles. And it is deeply sex-stereotyped. Girls and boys are expected to help with different tasks around the home; different occupations are considered suitable for them. Most parents would not oppose their child getting a job stereotypically associated with the opposite sex. But they do not envisage this happening. These everyday assumptions about the way their lives are organised, both now and in the future, probably have more influence on children than abstract and rarely expressed ideas about equality. When they are asked to think about the issues most parents endorse equality between the sexes. But when they are not thinking about it their actions are very different.

The practical implications of this for the school are mixed. Apart from a vociferous minority, parents will generally welcome all steps, such as mixed craft classes, towards formal equality; but they may be more hostile to measures, such as class discussions of sex roles, which try to challenge the basic organisation of society. Perhaps this hostility can be diffused by a careful process of parent education. One of the most striking features of our results was the very high aspirations parents have for their children, and the strong desire to support them in their own decisions. The confining influence of sex roles is largely unrecognised. But if parents are helped to see how traditional assumptions can limit the development of both girls and boys, their personal interest in their children's welfare may lead them to support destereotyping work. The foundations of sexual equality in parents' attitudes have been laid, but there is still a lot of building work to do.

The theoretical implications are also mixed, although this is not the place to explore them thoroughly. As Arnot (1981) has suggested, gender socialisation is not a uniform, unproblematic process. Parents' ideas and behaviours are often contradictory; the same is true at school where a formal ideology of equality often coexists with sex-stereotyped expectations from teachers. Moreover sex differences at school are not merely reproduced from home; they are reconstituted. At home sex differentiation takes the form of different household tasks and spare-time activities. At school this is transformed into different subject choices in a way which is by no means clear, but is certainly not a simple reflection of parental wishes. The occupational structure and children's expectations for their future may provide the missing link; but again the connection is not straightforward. At school children are differentiated by class and sex in the subject and level of study; on leaving school they are similarly differentiated in the occupational structure. Yet this study found a basic similarity between middle-class and working-class parents in their attitudes and behaviour towards girls and boys. Clearly a lot more work is needed to tease out the tangle of connections between sex, class and race at home, school and work.

ACKNOWLEDGEMENTS

We would like to express our gratitude to the Head and Deputy Head of Tall Trees School for their assistance with this research. We are also grateful to the parents who patiently answered our questions.

REFERENCES

Arnot, M. (1981) 'Culture and political economy: dual perspectives in the sociology of women's education', *Educational Analysis*, **3**, (1).

Borley, J. (1982) MA thesis, University of Manchester

Clarricoates, K. (1980) 'The importance of being Earnest...Emma...Tom ...Jane.' The perception and categorization of gender conformity and gender deviation in primary schools, in: Deem, R. (ed.) *Schooling for Women's Work*. London: Routledge & Kegan Paul.

Craft, M., Raynor, J. and Cohen, L. (eds) (1980) *Linking Home and School: a New Review*. London: Harper & Row.

David, M. (1980) *The State, Family and Education*. London: Routledge & Kegan Paul.

Delamont, S. (1981) *Sex roles and the School*. London: Methuen.

Kelly, A., Smail, B. and Whyte, J. (1981) *Initial GIST Survey: Results and Implications*. Mimeo, GIST, 9a Didsbury Park, Manchester Polytechnic.

Llewellyn, M. (1980) 'Studying girls at school: the implications of confusion', in: Deem, R. (ed.) *Schooling for Women's Work*. London: Routledge & Kegan Paul.

Maccoby, E.E. and Jacklin, C.N. (1975) *The Psychology of Sex Differences*. Oxford: Oxford University Press.

MacDonald, M. (1980) 'Socio-cultural reproduction and women's education', in: Deem, R. (ed.) *Schooling for Women's Work*. London: Routledge & Kegan Paul.

McRobbie, A. (1978) 'Working class girls and the culture of feminity', in: Women's Studies Group (eds) *Women Take Issue*. London: Hutchinson.
Newson, J., Newson, E., Richardson, D. and Scaife, J. (1978) Perspectives in sex-role stereotyping, in: Chetwynd, J. and Hartnett, O. (eds) *The Sex Role System*. London: Routledge & Kegan Paul.
Sharpe, S. (1976) *Just Like a Girl. How Girls Learn to be Women*. Harmondsworth: Penguin.
Smail, B., Whyte, J. and Kelly, A. (1982) 'Girls into science and technology: the first two years', *School Science Review*, **63**, 620–630.
Wolpe, A.M. (1977) *Some Processes in Sexist Education*. WRRC Publications.

8

*Girls and Boys in Primary Maths Books**

JEAN NORTHAM

SUMMARY

This chapter reports on an examination of a number of maths books originally undertaken as part of a contribution to a maths diploma course for primary teachers. The books cover the age range 3–13 and were published between 1970 and 1978. It was discovered that mathematical and scientific skills become increasingly defined as masculine as the pupils move through junior and middle school. Men feature predominantly in these materials throughout the age range and there are few models of adult females operating in the world of mathematics. The representation of girls becomes more marginal and they almost disappear from the scene towards the upper limit of the age range. The fading presence of girls in these books parallels the decline in girls' involvement and achievement in maths between infant school and GCE 'O' and 'A' level examinations.

As evidence has mounted on the differential achievement of girls and boys in mathematics and science there has not been the widespread concern with the portrayal of sex roles that was shown in relation to reading materials. The report of the National Child Development Study (Davie 1972), for example, indicates that by the age of seven, boys are marginally ahead of girls in certain mathematical skills. Whereas the report draws attention to the part

*This chapter first appeared in *Education*, **10**(1), Spring 1982, pp. 11–14. Reproduced with permission.

reading materials may play in the slower progress of boys in reading, no such suggestion is made in relation to girls and maths. Indeed, the problem arithmetic test included in the appendix to the book suggests that the researchers were not conscious that such a possibility might exist. Out of ten problems in the test, five mention people, namely Peter, a man, John, four boys and a boy.

Maths books do not usually present information about the social world in the coherent, narrative way that usually occurs in reading books. There is no story line, little characterisation, and experiences are explored for their mathematical properties rather than their intrinsic interest. However, children's test books are often lavishly illustrated and glimpses of social life are found in the problems and in the explanations of mathematical processes. There are notes for teachers which convey information about the assumptions authors make about children and childhood, and in schemes for younger children there may be guidelines on assessment and records of progress. The selection of areas of experience defined as mathematically interesting and relevant gives clues to underlying attitudes and values. When the style and content of the pictorial and written material in the books are analysed, it is possible to detect a particular view of the social world which forms a 'hidden curriculum' in the teaching of maths.

The books discussed in this article were not chosen as 'worst offenders' in their portrayal of girls and boys. They were selected from schemes known to teachers taking a maths diploma course, and cover a variety of styles and approaches. For younger children in nursery and early infant classes, the *Early Mathematics Experiences* (EME) (Schools Council 1978) and Nuffield *Mathematics — the First Three Years* (1970) were studied. Both series are addressed to teachers rather than children. *Mathematics!* books 1a, 1b, 2a, 3a were included as books addressed to children in infant and first schools (Golding 1971). The years 8–11 are covered by *Maths Adventure*, books 2, 3, 4, (Stanfield 1973) and *Discovering Mathematics*, book 2 (Shaw and Wright 1975) takes the study into the lower secondary or upper middle school years. Thus different publishers, styles of presentation and approaches to maths teaching were represented.

Women dominate the child's world in the infant teachers' books. Though unnamed children are invariably 'he', the teacher and the parent are 'she'. The Nuffield book describes the world of the under-five as with mother at home; fathers are rarely mentioned. The contents of mother's hand-bag form a set (lipstick, pen, purse, diary with pencil, nail scissors, powder compact), mother's necklace is an early object of interest, and she unpacks the shopping, takes the child to buy new clothes and does the cooking. EME chooses the word 'parent' for these early experiences but for the most part defines the under-five as a pupil in a nursery or infant class. Mathematical experiences are rooted in the provision for play that is made by teachers. The stereotyped picture of family roles is reinforced in both series by the presentation of the 'typical' family, consisting of father, mother, brother, sister and baby in descending order of height. In the book on the Home Corner, the EME team make this reservation: 'Usually, we assume that dad will be the tallest, mum the next and so on, but of course this may not be so and unless we know the family there may be little point in exploring the idea'. It is relevant to consider how far the model corresponds to the experience of most children in a number of respects besides that of relative height. Maths books tend to confine themselves to this model, however, throughout the primary age range.

There is little differentiation in the presentation of children engaged in play with basic materials and in sorting activities. In EME illustrations often show mopheaded trousered figures who could be either male or female. The dominant characteristic of children in all the infant books is not their gender but a particular kind of dependence. They are chubby, doll-like and expressionless apart from a vague smile. They sometimes lack sense organs such as noses, mouths and fingers, yet the clothes are carefully elaborated and decorated, the hair curled, beribboned, plaited or appealingly tousled. The arms and legs may lack joints, the feet are stiff and stubby. The child is presented as a passive receiver of adult attention, staring round-eyed out of the picture, physically and psychologically incapable of self-initiated action.

Two considerations underline the significance of the concept of childhood implicit in the three infant schemes studied. The first is the marked difference between the portrayal of children and women and that of men, which will be considered later. The second is the close correspondence between this concept and that implicit in some of the items in EME's suggested methods of recording.

In the EME guidelines we find the following items of behaviour used as criteria for assessment of developmental progress: 'helps with activities', 'keeps himself occupied', 'puts on own shoes', 'talks freely to strangers', 'understands simple instructions quickly' and 'draws writing patterns'. These items are not necessarily and directly related to the development of skills and abilities but may well be viewed as outcomes of adult/child relationships. The picture is one of the compliant, conforming child whose behaviour does not delay or baffle or challenge or trouble the teacher. It is a pattern which conforms to a female rather than male stereotype of behaviour and, if it accurately reflects teachers' expectations of pupils, seems designed to reinforce the kind of teacher-approved behaviour in girls which is likely to impede the development of self-assertion and initiative. Certainly, evidence from other sources suggests that infant teachers rate girls' behaviour more highly, on the whole, than boys' (King 1978) and this would not be surprising if their criteria resemble those suggested by EME.

As the analysis of books for older children shows, the concept of the pupil in infant books is at odds with the definition of the pupil-mathematician. What is particularly interesting is that it is also at odds with the descriptions of children given in the teachers' notes in the infant books. Nuffield, for example says: '...children must be set free to make their own discoveries and think for themselves'. A similar message is expressed in EME: 'The same learning experience goes on as they explore the whole gamut of toys, books, music and physical activity, gaining experience in the way materials must be manipulated'.

The active, exploring, discovering child of the teachers' introductory notes is similar to the pupil-mathematician in *Maths Adventure*, but may not emerge with too favourable a rating on drawing writing patterns, putting on shoes, or talking freely to strangers: any relationship between the two styles of behaviour would appear to be coincidental. It seems significant that Nuffield and EME are written for teachers, and therefore it must be concluded that the style of presentation is intended to appeal to them rather than the children.

Two further considerations related to infant maths are suggested by the examination of the *Mathematics!* series. These books are addressed to children, and are therefore, especially in the early books, more abundantly illustrated. One of the first sets the children are invited to make is that of girls and of boys. This suggestion is also made in the other infant books, but here the implications of this division become more clearly apparent. In order to distinguish clearly between the two sets, differences are emphasised. Dress, demeanour and sex-typical behaviour must be sharply contrasted so that the notion of 'set' is clarified. Thus we find no mopheaded, trousered boy/girl figures in the sets. All the girls are in dresses and have elaborate hairstyles; four out of five of them are standing gazing into space. The boys' hair is cropped short and out of seven boys, five are on the move. This example, taken from book 1b, further reinforces the differences by adding the instructions: 'Draw a chair for each girl', 'Draw a drum for each boy'.

The visual material contains a number of adult figures, cowboys, soldiers, sportsmen, clowns, pedestrians and tractor drivers; they are tall, short, fat, thin, they climb ladders, they mend roofs, and they are almost all men. Two examples only depict women. Mr and Mrs Smith are seen walking to church and two female heads are seen in relation to three hair ribbons. The men are highly individualised in the ways described and also in terms of facial expression and characteristics. Whereas the women appear bland, expressionless, and lacking in individuality, the men have ages, personalities, distinguishing features, occupations, tasks and intentions. Men are not depicted staring vacantly out of the picture, in sharp contrast to the portrayal of women and children.

Women disappear almost completely from the junior books studied. There are more than forty references to men, six to women. *Maths Adventure* adopts something approaching a narrative style, in that six major characters are followed throughout the series. These are Gary, Clarence, Joe, Jill, Ann and eccentric Uncle Harry. The initial impression on glancing through the book is that sex-role differences are somewhat diffused by the characterisations and by situations in which all the children are involved in the same or similar activities. In order to see whether this impression would bear closer examination, the vocabulary was analysed. Sentences which begin 'Gary noticed . . .' or 'Jill invented . . .' suggest particular skills; they imply in the first example, an ability to identify problems and in the second initiative and inventiveness. 'Ann recorded this way' suggests an elaboration of a process already learned. Instances of behaviour were categorised in the following ways:

(a) identification, setting and solving of problems;
(b) taught, explained processes to others;
(c) made something, displayed a skill;
(d) planned, initiated, invented;
(e) performed, played tricks, boasted;
(f) competed;
(g) repeated or elaborated upon a process already learned;
(h) co-operated, shared, helped, complied;
(i) corrected another's behaviour; e.g. 'Calm down', said Ann.

In Table 8.1, examples which involve both boys and girls have been omitted.

Table 8.1

	(a) Problems	(b) Teach	(c) Skills	(d) Initiate	(e) Perform	(f) Compete	(g) Repeat	(h) Co-op	(i) Correct
Boys	27	35	9	11	12	7	8	6	3
Girls	10	10	2	3	0	2	19	12	13

The initial impression, that girls are featured in proportion to their membership of the group, is substantially supported by this closer analysis. The roles they play in the group, however, are strikingly different from the boys'. They are featured as less likely to be involved in the identification, setting and solving of problems, less skilful and competitive, less likely to teach maths skills to others, and to display less initiative and inventiveness. Significantly, they are never shown performing in a play or boasting or playing jokes, activities which appear to be associated with self-assertiveness in the boys. They are efficient record keepers, they practise and modify already learned mathematical skills, develop themes suggested by others and set standards of behaviour. The girls continue to conform to the standards described in the EME books, while the boys develop interests and skills specifically related to discovery and exploration.

There are a number of less subtle examples of sex typing:

> Jill and Ann made machines which frightened even the boys.
> (Ann said) 'I am a fair damsel and when I drop this handkerchief you must fight for me'.
> 'Don't you know, Jill', said Joe, 'he is crazy about you'. Clarence blushed.

Food, pets and toys continue to feature prominently in problems, but the context is the peer group and outdoor life rather than the family at home. Maths in the junior books moves away from the domestic sphere and takes to the sportsfield, the battleground and worlds of business, space travel and machines where no women are to be found. In *Discovering Mathematics*, not only are women under-represented but girls have also disappeared from the content of problems and illustrations. In the 44 pictures with human figures, 29 are of males only, 3 of females only. Out of 109 figures, 68 are men, 21 women, 14 boys and 6 girls. The women are almost without exception confined to sitting in cars and buses driven by men, watching men at work or being rescued by a fireman. The men are using telescopes, making calculations, building houses, mending roads, they are shop-keepers, cricketers, skin divers, soldiers and waiters. One picture shows a male chef on television demonstrating cake baking and being copied by four women.

A similar picture emerges from an examination of the problems, in which boys are mentioned 23 times, men 18 times, girls 7 times, and women once. The impression that maths is and inevitably must be concerned with masculine activities is reinforced by descriptions of the mathematical prowess of people in history, illustrated by pictures of men and punctuated with references to famous mathematicians, Hipparcus, Napier, Pythagoras, Pascal. There are occasions when maths is distanced from childishness, 'softness' and the subjective content of situations by humour and by author's comments. A sentence opposite a picture of Goldilocks and the three bears reads:

> No! The picture opposite has not got into this book by mistake.

There is a clear tendency in the books studied to define mathematics as the province of males, and especially adult males. The fact that the social world is presented in fragments, through illustrations and problems, does not necessarily diminish its significance. Individual examples of behaviour, briefly and starkly presented, accumulate as one works through the books into highly stereotyped images of males and females with little of the blurring and elaboration that may occur in a longer narrative. The impact of such images is unlikely to be reduced merely because the social content is intended to be secondary to the mathematical.

Adult women are largely absent from these books, and by the age of thirteen girls have joined them in near-oblivion. There is an interesting parallel between the decline in girls' involvement in maths between seven and sixteen years of age, and the gradual disappearance of girls from maths books over the same period. It suggests that ameliorative action could profitably be focused on the values and assumptions which form the hidden curriculum of maths books and possibly that of teaching and learning in the classroom.

REFERENCES

Davie, R., Butler, N., Goldstein, H. (1972) *From Birth to Seven*. Harlow: Longman/National Childrens Bureau.

Golding, E.W., *et al.* (1971) *Mathematics!* (1a, 1b, 2a, 3a,). Aylesbury: Ginn.

King, R. (1978) *All Things Bright and Beautiful?* Chichester: John Wiley.

Nuffield/CEDO (1970) *Mathematics — the First Three Years*. Edinburgh: Chambers and Murray.

Schools Councils (1978) *Early Mathematical Experiences*. London: Addison Wesley.

Shaw, H.A. and Wright, F.E. (1975) *Discovering Mathematics*, (2) 3rd Edn. London: Edward Arnold.

Stanfield, J. (1973) *Maths Adventure* (2, 3, 4). London: Evans Bros.

9

*Boys Muscle In On The Keyboard**

MARY GRIBBIN

Microcomputers are promoted as the saviour of the nation. Both girls and boys, we are told, are provided with new hope for work by this new technology, and at Christmas 1983 the home computer was the present for the kids that walked off the shelves. The message hammered home by advertising was that a home computer was essential for children to keep up with what they were learning at school, and for parents to catch up on what they had missed. It clearly got through, with the implication that when it came to getting a job, with 4 million unemployed, that school leavers who knew the difference between a bit and a byte would have a head start. Even though a micro system could set parents back by hundreds of pounds, and shortage of the most popular models caused a frantic search for alternatives (often more expensive and less suitable), the micro was, indeed, the present that little Johnny got for Christmas 1983, and sales continue to boom.

But what of little Jill? There is a growing weight of evidence that girls are missing out on computer education, for all the reasons that they miss out on science education in general. And parents, worried enough about a son's job prospects to fork out hundreds of pounds for a micro in a desperate attempt to help, generally seem less determined to promote the job prospects of their daughters.

---

*This chapter first appeared in the *New Scientist*, 30 August 1984. Reproduced with permission.

The problem was brought home to me by the experience of children in a class of nine–ten year olds that I was teaching. All but one of the 26 children in the class said before Christmas that they would like a micro. Most of the children in the school came from fairly affluent homes, and 13 of the 26 children had their wish granted. Twelve of the chosen 13 were boys. Five of the girls who wanted a micro but didn't get one said that their brothers (not necessarily older brothers) had got one for Christmas, and that 'sometimes he lets me have a go'. And all of the children who did have micros—the few girls as well as the male majority—were taught how to use it by Dad. None of them ever mentioned Mum using it at all—which rather gives the lie to one current TV ad which shows Mum and Dad happily playing with the new toy while little Johnny looks on. Typically, I was told by my pupils, Dad 'helps' the children with the computer in the early evening 'while Mum is cooking the dinner', or at weekends 'when Mum is clearing up the house'. At the age of nine, children in our society are conditioned to accept that boys and men are the proper users of a computer, that girls might be allowed an occasional touch of the keyboard, and that a woman's job is to feed and care for the men.

I thought that this might be a purely local phenomenon, and certainly would not have dreamed of drawing far-reaching conclusions about the state of the nation from such a casual and limited 'survey'. But now the London Borough of Croydon has produced a report for the Equal Opportunities Commission (EOC) on *Information Technology in Schools*. These 'guidelines of good practice for teachers of IT' (the subtitle of the report) deal with the school environment, not the home. But they report exactly the same sexual stereotyping at work, and identify a key problem as the natural tendency for a school micro to become the property of the maths department, which is dominated by male teachers (a legacy of previous generations of sexual stereotyping) and to be seen as 'out of bounds' for girls.

The EOC report includes the results of a survey of 1200 children, roughly evenly split between boys and girls, from a mixed catchment area in south London. Out of 109 children who had access to a computer at home, only 30 were girls. Children between the ages of 5 and 11 were asked what they thought a school computer would look like, what it would be able to do and whether it would be more useful for boys or girls or both. First-year juniors — eight-year-olds — were open-minded and expected both girls and boys to find the computer useful. The only concern about restricting its use was the opinion voiced by one or two of the pupils that smaller children should be kept off for fear that they would 'muck it up'. But already in the second-year group, a few boys were expressing the view that girls would have more difficulty in getting the computer to work. And at this point differences in attitudes began to emerge between children from different socioeconomic backgrounds. Children from a school in a middle-class area tended to be more aware of the importance of the computer in a changing technological world, and to perceive little difference in the role of men and women. Most of them thought that computers would be used equally by all; comments included 'I think the computer is for boys and girls because you can use it for lots of things' and 'it would be a lot more interesting than the teacher, writing on a blackboard'. Just a few boys suggested that the computer would be 'for boys mostly, though girls can use it'.

The responses were different when the same questions were asked of children in a school in a predominantly working-class area. Many of these children were disadvantaged in some way, and clear signs of chauvinistic attitudes were appearing even among the second-year juniors. Both girls and boys thought that 'boys should use computers', and among the third and fourth years comments included 'boys would use it more because girls would go out shopping all the time' and 'the computer can help you with any research . . . girls couldn't work it'. One girl said: 'I think the computer is for boys and girls. The boys say that it's not fair, but I think it should be for girls because boys get everything.' She could probably have done without the support of the one boy who agreed that 'girls will be far better at using the computer because girls are better at typing'.

This highlights a major problem with the present system which has strongly encouraged primary schools to invest in a microcomputer by providing half of the cost from a DES grant. And this problem is aggravated by the folly of putting the cart before the horse in getting the micros into the schools without having qualified teachers who understand the educational potential of the machines and have been trained to make best use of them in a wide variety of subject areas. The schools are happy because they have got money out of the DES (but how happier many of them would be if they could have the same amount of money to spend on pencils, which are in embarrassingly short supply in some schools); the PTA is happy because it seems that the school is moving with the times.

But as often as not the computer sits in its box unused for much of the time, or is wheeled out so that the children who know about them from home (usually the boys), may, perhaps with the aid of a volunteer parent (usually a man) show some of the other kids (seldom the girls) how to play games. Croydon is one of the most alert local education authorities (LEAs) in the country when it comes to awareness of the difficulties of using IT in education, and runs several in-service courses for primary teachers to learn how best to utilise their school micro. These courses are so vastly oversubscribed that a school which puts a teacher's name forward for one today may have to wait six months before he can go on it—and I use the term, 'he' advisedly. Robin Ward, who is now a project worker jointly funded by the EOC and Croydon and was a major contributor, as an English teacher, to the EOC report, told me that it is almost always the school's maths/science specialist who is given the task of finding out about the micro, and among the majority of women teachers in primary schools the one post which is most likely to be held by a man is that of the maths/science specialist. The result is that although the vast majority of primary school teachers are women, 40 per cent of the ones who attend one of the Croydon IT courses are male—science specialists, or head teachers who put themselves forward for what may be seen as a perk.

Clearly, any attempt to encourage more girls into science in the 1980s and 1990s must begin in the primary schools, where the first 'scientific' piece of machinery the children encounter is now certain, thanks to the recent government programme, to be a computer. The present *ad hoc* arrangements are, if anything, providing yet another, very early hurdle for girls to overcome if they are ever to become scientists. By the time the present juniors are in secondary school and beginning to look towards 'O' levels or their equivalent, lack of computer expertise may itself be a reason to turn girls away from scientific subjects. So what can be done to redress the balance, and to take advantage of the opportunity to show that girls are just as good at working with computers as boys, perhaps even providing a basis later on in their educational careers to argue that since they *are* good at computing, that is a good reason to allow/encourage them to do more science?

Many of the conclusions in the EOC report seem no more than common sense. Yet it is essential that they are heeded: common sense has been conspicuously lacking in the ill-conceived rush to get computers into schools. 'The Information Technology teacher,' says the report, 'does not need to be a computing expert, but rather be a good communicator with an interest in the technology.' LEAs should have a duty to provide properly equipped and safely wired rooms for computer work, in the same way that science, craft and home economics rooms are provided (not, as I have experienced, a loose cable trailing across the room to a three-way plug and a computer system balanced on a rickety desk). A coherent and comprehensive programme of training teachers is essential, and seminars should be organised to inform head teachers of the role of these trained teachers.

Damningly, the report says that 'there is a dearth of good teaching materials and a lack of proven software', for pupils and teachers of both sexes to use; and, concerned about the rapidly developing image of the school computer as primarily for the boys, it urges that the LEA should obtain statistics from schools on the relative popularity of different subjects for boys and girls, and their relative success in examinations. And, as well as offering specific advice to schools on how to make best use of the machinery dumped in their laps by the DES, it stresses that girls should not be discriminated against, for example, by setting up subject options which make it difficult for a girl to choose computer studies ('sorry, Jill, that clashes with Domestic Science'), by allowing unsupervised group or club activities where aggressive boys muscle the girls out, or by undue linking of IT with what are perceived as boys' activities.

A collection of minor points and hints, including advice not to put one girl in a group with several boys since 'girls will assert themselves much more when other girls are present to support them', add up to a powerful indictment of the present system into which we have, in the great British tradition, just drifted. They also chart the best way out of the mess more clearly than any other currently available publication. The EOC has already sent a package including the report, a computer comic called *Load Runner* in which boys and girls share equal status, a leaflet showing job opportunities for girls, a booklet *Working with Computers* and a newsletter from F International, a systems company run by women, to every secondary school in the country.

10

Characteristics, Views and Relationships in the Classroom

ROSIE WALDEN and VALERIE WALKERDINE

INTRODUCTION

This chapter is taken from the larger work (Walden and Walkerdine 1985) which examines the transition from the top of the primary school to the bottom of the secondary school of a group of girls and a small group of boys. We were interested in how children cope with this transition and with different school practices when mathematics becomes a more discreet area of the curriculum in the secondary school. Our particular concern was that girls, whose performance at the top of the primary school is considered good, show a decline in performance by public exam time.

The study used a variety of methods. These are detailed in Walden and Walkerdine (1985) but briefly consist of tests of mathematical attainment, repertory grids, with the children, interviews with teachers and children and observations of classrooms using video tapes.

We have chosen to present here the chapter which deals with the observational data for specific reasons. Prior to our work, research in this area in Britain had often, but not always, tended to focus on large-scale surveys of attainment. It is true that the American literature has concentrated less on the attainment differences and more on girls' failure to choose mathematics courses, using theories of motivation and gender identity (see Diener and Dweck 1978, p.451, 462, Bar-tai 1978, p.48). In Britain researchers in this field have tended to draw from large-scale surveys certain inferences about sex differences. We were concerned that there was some reductionism in these approaches, which were not able to engage with how gender (the social aspects of femininity and masculinity) is produced in the classroom.

Another of our concerns was that conclusions drawn from large-scale surveys, and considered to be significant, begged several questions. The issue was not so much about whether there were sex differences, but rather how big these differences were and how important they proved to be in practice. Conclusions drawn from large-scale surveys tend to account for differences between populations (in this case girls and boys) by the application of tests of statistical signficance. We are wary about some of the inferences drawn from such work because the issue of statistical significance is the object of major debate within social scientific research (see Carver 1978, p. 378–399, Morrison and Henkle 1970).

The term 'significant' applied to such results gives them a legitimacy that can be over-rated. If we look at the difference in distribution of girls' and boys' scores, as well as the difference in their mean scores, it is the case that the bigger variance for boys in many areas is well known. (This was made much of by, for example, Corrinne Hutt.) In our Bedford Way Paper (1985, p.29–30) we pointed out that two of the Assessment of Performance Unit surveys have found that the main difference between girls and boys was the greater preponderance of boys among the high scorers. We have found much the same in our own data. However, we feel that we must make the point that arguments for treating girls and boys differently have not been based on just the top 10 per cent of scorers, or even the top 25 per cent. Rather they have been based on 'all' girls (as in the sex-linked spatial abilities literature) or for girls 'on average' (see Scott-Hodgetts in this volume). We would seek to counter these arguments by demonstrating that not all girls are bad at mathematics, and, more importantly, that there is very little difference between girls' and boys' performance. Even the APU shows that.

Although it is standard in the literature to make extrapolations about classroom performance and ability from test data such as these, we suggest that inferences are often made much too starkly. The differences in performance between schools and tutor sets in our data (and between different areas and poor and wealthy schools found by the APU) lead us to the supposition that the social relations, teaching and learning in specific classrooms, are of considerable importance in accounting for test attainment.

It is for these reasons that we have chosen to reproduce here one of the chapters in which we demonstrate the importance of classroom practices both on girls' performance and the evaluation of their performance by teachers.

THE CLASSROOM (CHAPTER SIX)*

We have argued that there is a particular combination of classroom practices and an understanding of mathematical learning which produces failure in girls, and that in consequence girls are put into the position of being successful but not succeeding. Here we shall examine some of the parameters of the classroom production of this situation for girls, illustrating with examples from our own study. Since the amount of data, case material and observation is vast, we can provide only a glimpse of the kind of analysis which we have carried out. We have decided that the most effective way to present it is to set out some of our more important analytic categories and then to illustrate these with reference to the case of particular children (girls and boys) who represent specific polarities in the positionings we describe.

In both primary and secondary classrooms our fieldwork consisted of fieldnotes, observation of classroom practices and videotapes of the specific performance and interaction of the children in our sample. Each of these children was videotaped for one hour. Together, the fieldnotes, test interview and grid data, and videotapes, form a massive amount of information on children and classrooms. Our discussion here will concentrate on the presentation of fragments from our transcripts illustrative of particular concepts derived from our analytic framework. We shall begin by elaborating what these are.

Our pilot work (Walden (Eynard) and Walkerdine 1981, 1982) identified certain concerns which we felt offered potential explanatory devices for the apparent phenomenon of discontinuity. As we have said, we have not discovered a simple discontinuity as such, rather continuities in the data which support our previous work on early success, and make of later performance much more of a problem than was previously envisaged.

---

*This chapter first appeared in Walden, R. and Walkerdine, V. (1985) *Girls and Mathematics: From Primary to Secondary Schooling*, Bedford Way Paper. 24, University of London Institute of Education. Reproduced with permission.

Let us set out in detail some of the categories we used and have continued to develop in relation to this data.

Power, positioning and gender

We have argued strongly against an analysis which understands girls as powerless because feminine and against a model of girls simply 'squeezed out' of academic performance. Relations of dominance and subordination, and of power and resistance, can be explored in relation to the social relations of the classroom. We shall argue that girls are not subordinated in any simple or once-and-for-all way but that they can move from powerful to powerless positions from one moment to the next. Femininity and academic achievement are in this analysis not incompatible, but their relationship, as we have been trying to show, is neither without *problems* nor without specific *effects*.

Our analysis depends upon a theoretical framework elaborated elsewhere (for example, Walkerdine 1981, Henriques *et al.* 1984). However, for our purposes here it is important to state that we are critical of a monolithic view of power which understands it as a unitary possession vested in particular individuals simply by virtue of their institutional location. By such analyses power is invested in teachers and boys and not in girls. While we would not wish to argue that the relations in terms of institutional position of teacher and pupil, boy and girl, are ones of simple *equality*, what we are addressing are the ways in which relations of power and powerlessness are produced in classrooms and the forms and content of understanding available to teachers which operate in educational practices. In their formal sense these can be described as power and knowledge relations (see for example, Henriques *et al.* 1984) Such an analysis implies criticism of some formulations of the sociology of knowledge (for example, Young 1971) in which knowledge is powerful because it is *possessed* by a particular group (such as teachers).

A particular site or positioning which, as our pilot work demonstrated, allows girls to be powerful in the primary classrooms is the position of *sub-teacher*. By being positioned like the teacher and sharing her authority, girls are enabled to be both *feminine* and *clever*; it gives them considerable kudos and helps their attainment. In a variety of ways the *relations* of power and powerlessness, helping and being helped, may be shown as existing between teachers and children and between children. Some girls will be helped by one set of children and be helpers to another, powerful in one set of relations, powerless in another. For example, a girl may be popular and not academically good. By examining those practices and contexts which we outlined in the last chapter we can see that there are relative (and to some extent cumulative) powers. A girl who is located as powerful in helping and in sport, for example, has a very high status with other children. However, she may still not be considered to have 'flair', to be 'really' or 'naturally' bright by the teacher. There is, then, a difference between the pupils' and teacher's estimations, and, this in turn relates to the issue of femininity and classroom attainment. They may be complementary or even contradictory. These different positionings produce and affect different girls differently, but their total effect helps produce the possibility and reading of attainment in the classroom.

Although the positioning of mathematical success in terms of *rule following and challenging* interacts with the masculinity and femininity dimensions we have argued against a simple reading of the production of independence and autonomy in girls by making them more like boys as a prerequisite to academic success. What we shall explore is the *effect* of a certain kind of confidence in rule-challenging procedures in the evaluation of performance by the teacher. We argued in our early work on primary school mathematics (Corran and Walkerdine 1981) that the practices of mathematics teaching relate to procedural rules. These rules are both behavioural and organisational in relation to the classroom, and *internal* to the organisation of mathematical knowledge itself. In order to be successful children must follow the procedural rules. However, breaking set is perceived by teachers as the challenging of procedural rules internal to the organisation of mathematical knowledge itself. It is read as 'natural flair'. In the first instance naughtiness (most often displayed in boys), i.e. breaking behavioural rules, can be taken as evidence of a willingness to break set, to be divergent. Consequently girls' good behaviour is taken to be evidence of passivity, rule-following, and hard work. Later, however, bad behaviour in class tends to be expressed as anti-intellectualism which can no longer be read as playful but as oppositional (Walkerdine 1981, Willis 1977).

To challenge the rules of mathematical discourse is to challenge the authority of the teacher in a way which is sanctioned. Rule following and rule breaking are both received forms of behaviour even though they are antithetical to each other. If there are considerable pressures specifically on girls to behave well and responsibly, and to work hard, it may well prove more than they can bear to break rules. Firstly, they would risk exclusion by others for naughtiness and secondly they would require the confidence to challenge the teacher. Such contradictions place them in a difficult, if not impossible, position. Our analysis leads to the conclusion that there are problems for girls, but *not* of the order of a fundamental or essential lack of ability, or of missing out a natural sequence of correct development towards cognitive maturity. Rather, social and psychic relations coalesce to produce possibilities, positions and constraints which both allow and prevent certain forms of behaviour and of thinking.

For example, to understand the contradictions involved in rule breaking and the problems attached to speaking out (see Spender and Sarah 1980) is very different from an analysis which suggests girls have simply either 'got something missing' or have been forced out, i.e. by a patriarchal conspiracy, currently the most favoured forms of explanation.

In the fragments of case material which follow we shall present examples of the kinds of positionings which we have outlined.

Individual cases: primary data

We shall begin by focusing on Patricia—a girl in school J1 chosen by her primary teacher as poor at mathematics. We shall see how she is produced and maintained as helpless. She was one of a group who were considered all to be 'much of a muchness'. She herself chose as her ideal self a girl whom she considered, ' . . . just like me really, not very good at things'.

The transcripts show her, along with her classmates, involved in doing an exercise on numbers. They have been asked to add all the numbers from 1 to 5, then 1 to 10, 1 to 15, and so on. Consonant with the practices in the primary school the teacher never makes explicit what it is he wants them to do. First they do some examples, then if it is not obvious to them he will explain the proposition he wishes them to understand. Eventually he talks of triangular numbers which is what the exercises have been leading to, but at this point the class are being given practice at manipulating numbers.

In Patricia's case it is apparent that she has misunderstood the teacher's instructions to add all the numbers. Her friends have to help her and she becomes steadily more anxious throughout the tapes. At the beginning of the second tape she tries to get the teacher's attention to see whether or not she is doing the correct thing:

> *What do I have to add it to, Sir?* (to Jo her friend) *He's just shown me how to do it, I'll find out, I've gotta add it haven't I? Sir, is this right? I've got it I think . . . Here you are, look here, here's the answer I got, Sir? Sir? Is this right?'*

The teacher ignores her to talk to another child. Patricia continues to try and attract his attention:

> *Sir? Is this right? I'll go and have a look 'cos if . . . if he's showing us how to do it I'll go and have a look . . . Sir, is this right? . . . Sir, I can't do these.*

The teacher's response, on looking at her work is to suggest that the work is too difficult and that she should do some easier work, which has been written on the board and explicitly marked as easy. This is an attempt on his part to help her—work she can do may not leave her feeling as demoralised as she obviously is at present. But all it succeeds in doing is confirming for Patricia her position as not very good, because it offers her security in her helplessness. Her position as powerless is underlined by each of the other dimensions mentioned earlier. At her school the teacher felt unable to intervene with Patricia other than to give her more practice on lower-level work which he hoped would help. Patricia's problem in attracting his attention meant she had to rely on her friends for help. The following exchange is typical of Patricia's contact with her friends. Patricia constantly requests her neighbours Ann and Jo to help:

> *J* (to Ann) *She's done it wrong.*
> *A* *Who?*
> *J* *Patricia.*
> *P* *Well, how do you do it? . . .*
> *A* *You're still doing it wrong.*
> *P* (to Jo) *Am I doing it wrong?*
> *J* *Yeah.*
> *P* *Why?*
> *A* *'Cos you're not copying the board.*
> *J* *You're not meant to copy the board. You're meant to work it out for yourself.*
> *A* *You're not.*
> *J* *You're meant to work it out yourself . . .*
> *P* *One to three equals four, right?* (she looks at Jo's book), *Yeah, Oh God, talk about thick* (she rubs out her work).
> *J* *You're doing multiplication. It's add.*

In fact Patricia has been adding the numbers but has failed to understand that the task is to add *all* the numbers from one to three not to just add one and three. In this episode Jo establishes her superiority in several ways: by talking about Patricia to a third party whilst ignoring her; she offers no advice on how to correct the problem, merely stating that she knows what to do whereas Patricia obviously does not. Ann's response to Patricia is also dismissed since the children 'know' that within the pedagogic framework of their lessons learning is achieved by doing the work not by copying it. Jo is able to dismiss Ann and deliver the *coup de grâce* at the end by nonchalantly 'spotting' Patricia's problem. From this point on in the tape Patricia asks Jo for help eight times.

Patricia is, therefore, always put in a position of being the one who is helped. Jo is dominant in the exchange and understands both the immediate problem and the rules of the classroom and acts as a sub-teacher to Patricia (and to Ann).

Let us take an example of a good girl in the primary school. Elizabeth is in a different position. At the same primary school as Patricia (J1) she was considered good by the teacher: 'She's interested and she finds it fascinating and she's interested in the odd bits of it.'

She liked mathematics when asked and considered herself clever. Unlike Patricia, Elizabeth rarely asked others for help and tended to work independently. In the lesson we taped on her she was getting bored with the work set and couldn't work out what to do:

> *. . . still came out wrong. I'm tired of this. I'm not doing it any more if I get it one more time wrong . . .*

Immediately after this the teacher approaches and singles her out for praise:

> *Children look at this neat and very good work. I want work like that . . . I want work at that standard.*

Without trying, Elizabeth has been able to attract the teacher's attention and at a particularly crucial time; it stopped her from becoming disheartened (she continued the task) and reaffirmed her position as a good child who did her work in the 'correct' way. This made her powerful in the social relations of the classroom. All the others considered her good at her work and she was sought out for help even though she only gave it unwillingly.

Thus, the practices of the classroom situate each of these girls in quite different positions, Patricia as powerless and Elizabeth as powerful. Moreover, her positioning helps Elizabeth to deal with problems. Even when she does not get her work right, her problems are dealt with in an opposite way to those of Patricia. She is treated by the teacher as able, and therefore knows that if she tries she will succeed.

The helping/sub-teacher positions occurred in both primary schools J1 and J2, but were the results of different practices used to teach mathematics. At school J2 the children were organised into groups to do specific pieces of work while at J1 the whole class most of the time worked on tasks together. At J2 the children were organised into groups to talk to and help each other and specific children were told to help others. For example, Sally was told to help Mollie:

> T *Now Sally, . . . that page is a page that you did very well. Do you remember organizing that? I'll give you a chance to organize with her, all right?*

The teacher returns later to make sure all is well:

> *So you understand what you're doing? Do you? You don't. Well, Sally, I thought you'd explained it to her. All right, you explain it again and I'll sit in on what you're saying.*

Later, the teacher continues to discuss the question which is about sets, but as the conversation continues her exchanges, although nominally with Molly, are actually with Sally.

> T *. . . So here you are, you had twenty-four rows of bushes. I want your help* (to Sally). *You've got twenty-four rows of bushes and you set them out . . . how? Is that how you set them out? Is she copying this example or not?*
> S *No Miss, she's not.*

One of the examples from this school involves two boys helping each other. This next exchange whilst partly humourous is indicative of how George approached his task of helping Ray. The task is to find the area of three pieces of carpet. First they have to estimate it, then measure it, and compare the two sets of figures.

> G *Right, just tell me what length is?*
> R *From there to there (touches top two corners)*
> G *From there to there (top to bottom corners of one side)*
> R *Correct (Stewart pretends to 'nut' him). What's area?*
> G *Area? The whole.*
> R *Right . . . you're learning.*

It also expresses how Ray also accepted the parameters being used both to define himself and George and to define the way of working. We include this example to show that these positions are not *essential possessions* of boys or girls. They may relate to the production of femininity and masculinity which means they have different *effects* when displayed in the two sexes. However, it is important to show that they can, and indeed do, overlap.

Continuities and discontinuities

Let us look now at these same children in their first year of secondary school. In the chapter on teacher interviews and in the opening chapter we pointed to the differences between the sectors of the educational system. Essentially, apart from size, the main difference is in the backgrounds and framing of the staff, most of whom are subject specialists with different views on teaching and learning. As it turned out most of our sample had two mathematics teachers splitting three periods a week in a 2:1 ratio. Thus another dimension was introduced—the differences between teachers in their readings of the children's performances.

Patricia's position at secondary school did not change much. Both teachers (Mr G and Mr H) considered her weak and one repeated the prescription from primary school that he saw her as part of a group considered 'much of a muchness'. She still had difficulty attracting attention and was often dispossessed of her place in the queue by more aggressive or dominant children. The taped lesson was about learning braille and attempting mathematically to discover the combinations of dots possible. By the end of the first tape the mathematics teacher came to see Patricia and again explained the task to her, to which her only response was 'Yeah'. As in the primary school she was constantly asking her friends what to do: 'Can you do that one?' 'Tell me what to do.' The teacher saw her less than any of the others and when he did so it was for a shorter period of time. She rarely questioned him and rarely seemed to have got what she wanted from the exchange. Mostly she sat quietly—one of the teachers called her 'non-noticeable'—her only discussions with the others being about her small nephew or to ask for help.

For George (from School J2) the pattern had changed, and his position was reversed. At this stage he was the one in need of help. SMILE work as operated by his teacher meant mostly individual work. George's only sustained contact throughout the lesson taped was with another boy, John. John's sub-teacher relationship to George had developed as a result of the way in which the work was structured. George was constantly questioning him:

> *What page did you start from? . . . Have you done these? . . . What's that one?*

Throughout the tape John helps George but in a way which affirms his superiority and George's inferiority:

> *. . . you're gonna have to do it, you're that stupid.*

This relationship, between the two boys, shows clearly the assymetricality of the sub-teacher/helper relationship. George's position as the *helper* in the primary school derived from the teacher's power in setting up the relationship; as did his position as the *helped* in the secondary school.

At secondary school Elizabeth was considered: 'obviously fairly competent' by one of the teachers and by the other as ' . . . confident . . . keen and eager. She wants to learn and she knows it takes some work on her part.' Elizabeth finds it easy to command attention from both her teachers. The rules of the classroom run by Ms R were fairly explicit in that children were praised for concentration—and told off for lack of it. They were also praised for good work, although the teacher often said that she didn't care about neatness but was more concerned with understanding what was written in books. Idle chat and gossip were frowned on, as was aggression. The girls were often told off for being trivial.

Elizabeth appeared to have none of these faults, although she was often aggressive to her neighbours if they could not follow her explanations or were being silly. She was easily discouraged when work was not going well (as we saw when she was at primary school) and claimed to be bored by the easiness of the work.

Ms R told the children that what they produced was the measure of what they could do and Elizabeth was often 'fed up' if she couldn't perform as she wished. In Mr P's lessons Elizabeth more often claimed to be bored—especially during SMILE lessons. Her confidence was manifest on the occasions observed when she was able to approach the teacher and demand a specific card be put on her matrix. Again, she helped others with their problems and was sought out for this purpose, even though she was often not very welcoming and could be very rude. She seemed to prefer Ms R's lessons, taking a dominant role, answering and asking questions and putting her head down and seeming to concentrate hard on the matter in hand.

If Elizabeth needed help she was able to obtain it almost instantaneously by demanding attention, and she often adopted the strategy (most favoured by boys) of just calling out answers. Elizabeth acted in what could be termed a 'masculine' way in that she was active, even aggressive, participating in all the classroom interactions. She was confident, especially about mathematics: 'I like it, I enjoy it. I can do it.' Although stylistically she acted in ways which were most often used by boys she did not want to be like a boy, nor did she like boys, choosing them in both primary and secondary school as people she most disliked. A masculine positioning with respect to academic practice does not, therefore, mean a wholesale or unitary masculinity. Her success had to lie precisely in managing a 'balancing act' of a 'masculine' placing in relation to academic work and feminine one in respect of helpfulness and non-work contexts, which ensured a stable feminine position. This is neither necessarily easy nor successful for many children.

Following and challenging rules

Let us look now at one of the areas we discussed earlier—the problem of following or breaking the rules. As we have said, it was considered important to the learning of mathematics to be able to 'break set', to 'free' oneself from the confines of simply following rules or learning by rote in order to discover for oneself. Our teacher interviews—both primary and secondary—are full of the distinction between mechanical and creative work, between flair, or natural ability, and hard work. The following examples will serve to show how certain types of behaviour were 'read' by teachers as exemplifying these dichotomies.

In the lesson taken by Mr H in the secondary school about codes and specifically braille, the differences between Patricia's position and that of another girl Charlotte exemplifies some of the points made.

He begins by asking them how many combinations of dots they could find, given that they could use any combination of up to six dots. Charlotte volunteers an answer and so does Jim, his answer depending on leaving one block blank with no dots at all. The teacher agrees that this is a possibility and is immediately challenged by Charlotte: 'You said to count the dots'. The teacher obviously doesn't want to get into too much of a discussion and consequently uses the device of shifting mathematics out of the 'real' world into a plane of its own by replying: ' . . . I think it's just as good maths if you count this one or you don't as long as you made it clear what you're doing'. So, 'in maths' anything goes as long as there's a reason for it. Charlotte, however, does not let things slip: twice she says: ' . . . we were talking about dots'. She is implicitly questioning his definition and insisting on her own. The teacher ends the discussion—which he can do, holding as he does the ultimate authority—by moving on to something else, pretending that her query is a non-problem: ' . . . I'm quite happy for you to take that attitude. I'm not arguing with you . . .' Being in charge and setting the parameters for the lesson, in that he is the only one who knows what is going to happen, he moves on. The teacher again explains the task and that braille is for blind people and consists of raised dots on a rectangle. Since the teacher began to discuss the task Patricia has kept quiet.

Charlotte is able to challenge the teacher and attempt to assert her own definitions of what counts in the lessons. She is very demanding of the teacher's attention, using a variety of tricks and strategies to obtain it. For example, she never queues, always by-passing the others and seeing the teacher immediately. Patricia on the other hand queues patiently for long stretches and is often passed over two or three times in favour of other, more aggressive children—often girls. Charlotte is very confident in class lessons calling out answers, questioning and challenging the authority of the teacher. On two separate occasions she was observed berating her teacher about SMILE, which she did not consider 'proper' mathematics. Not all of the attention she receives is positive—far from it, she is constantly told off for talking and not working hard enough. But this indicates that the teacher is constantly monitoring her progress. It is interesting to examine what the two teachers of this class, Mr G and Mr H, have to say about Charlotte and Patricia and how that fits with a notion of following and challenging rules as essential to the 'correct' learning of mathematics.

It is Charlotte who generates the most interest. Charlotte, for Mr G 'is the one with the keenest brain in the sense of ideas. And she's the great problem solver'. Charlotte is discontented with SMILE which she has told her teachers she thinks makes them lazy ' . . . because it's an easily manageable course . . . all you do is come in and the kids mark their own work . . . ' The idea of the children having to do all the work she finds wrong: 'She's constantly trying out ideas all the time and that's why she finds SMILE a bit of a constraint.'

He recognises her faults—selfishness, being mercenary in her use of her friends—and yet ' . . . she's a tremendous abstract thinker, she's great at the maths. That, perhaps, we don't recognise enough.' He considers he might be at fault in not using the matrices imaginatively enough. Overall Mr G believes in 'natural' ability and flair but is unable to be more specific. Nor is it clear whether he considers the cleverest children those who are most noticeable or those who cause him no trouble vis-â-vis his teaching technique.

Mr H felt that there were three children who 'really stood out' as good, two boys and a girl. He, too, felt that the good children were not of a piece and were not a recognisable group. He felt the girl to be 'a really outstanding mathematician', 'quite a phenomenon' who could do anything she was given. The boys were seen as confident and flexible. Another good girl he described as 'sharp . . . and quick at analysing a situation'. Finally he was able to sum up his feelings about all four of them that 'They have a certain sort of confidence.' It was this confidence which distinguished them: ' . . . they're not afraid of getting something wrong . . . They're prepared to try an answer, because if it's wrong then they'll just try the next one, or another approach.'

With regard to Charlotte, Mr H saw her as being better verbally than her written work had led him to suppose. He was impressed with her work during the lesson on braille—surprised even, since 'She hasn't done a great deal of work. She's always hard to get to do work . . . what made me think was that there is a sort of mathematical thinking going on in (her) which has never been tapped by me.'

What about Patricia? Both teachers, as we said earlier, saw her in the bottom four of the class. Mr G classed her, along with the other poor children, as ' . . . just finding it difficult to understand things'. He thought that Patricia and her friends would concentrate and work through something, although it worried him that they wanted to ' . . . get the matrix finished and carry on with another one' regardless of whether or not they had understood the work. He saw this as a particular problem for the poor children and a defect of the SMILE system. As a strategy for dealing with what seems to him to be a lack of confidence in the poor children Mr G sometimes set children like Patricia and her friends the same cards or their matrices: ' . . . in many ways to give them a little bit of security if they're sharing the same problem'. This tactic also expresses a wider concern on this teacher's part that the three girls, Patricia, Jane and Sue, are in danger of becoming disaffected with school for the reason which he sees as: ' . . . the big frustration thing. Not being able to do the tasks . . . I'm not trying to stretch any of them . . . I'm just trying to make them feel confident.'

Mr H simply sees her as ' . . . a very non-noticeable pupil . . . She sits very quietly, on the whole getting on with things, not asking very much. I think she's not confident . . . she's slow really . . . on the whole doesn't search me out at all.' Of course Patrica does try to 'search him out' but is often discouraged in her attempts and therefore withdraws, whereas Charlotte by her confident persistence and refusal to be discouraged attracts attention without any problem. On the other hand, Ray is at least as disruptive as Charlotte, yet his 'rule challenging' is not sanctioned and is considered disreputable and certainly not seen as evidence of 'correct' learning. He is always in the middle of noise and disruption and disturbs other children a lot. Unlike Charlotte who is offered various inducements to work (like different work cards, to keep her interested) Ray is simply admonished and told to get on. His form of challenging, shouting, laughing, etc., is less acceptable than Charlotte's whose challenges to the teacher's authority are on an intellectual level.

One more example from a different class will suffice here. We shall compare Helen, one of our sample children, with another girl in her class, Kay, considered by her teachers, Messrs. J and K, a good girl.

Mr J said: '(Kay) registers most with me . . . because she does have a very precise interest. She likes to get things straight, she's almost officious about it.' He thought she was one of the few children able to make 'connections between the different areas of their work'. Mr K thought her an ideal pupil in that she could 'combine presentation with a thorough understanding of the mathematical ideas and in some ways that's an ideal combination'.

It was in the mention of presentation that we see a clear dichotomy between rule challenging and rule following which also counterposes girls and boys. For example Mr K's best children were able ' . . . to organise . . . write down their work systematically . . . so that they can make sense of what follows'. However, later he categorises his poor girls, of whom Helen is one, as having a 'tendency to over-accentuate their presentation' in which he feels 'all that kind of detailed work . . . hides the main mathematical concepts'.

So there is an immediate contradiction between the importance of presentation and challenging the rules, even though the former is considered an important part of mathematics. A good boy did not ' . . . bother about his presentation, but I can see that he understands the mathematical ideas' and so Mr K chooses (as he himself says later) not to complain about the untidiness of his work in case it hinders his progress. This teacher's stress on presentation is unacknowledged and thus creates a double message and also a double bind for those who wish to do what he wants. Helen was considered not only to over-accentuate her presentation but also to be ' . . . very well mannered and polite and rather than push herself forward to understand more mathematics she'll sit back . . . ' When asked, Helen had this to say about mathematics: (They) 'try to explain . . . but there's always someone round . . . waiting and I feel I've got to hurry and I get in a mess.' Helen's diffidence and uncertainty caused her problems and she was the only person interviewed who mentioned being bullied. Other children in her class also mentioned that she was bullied. Kay, on the other hand, was seen very positively by all the children. She was the only first-year child out of the thirty-two interviewed who was successfully able to bridge the sex barrier and be accepted by boys and girls—on their terms. She was able to partake in masculine activities, particularly football, and still be seen as feminine as this quote from Ruth shows: 'She enjoys a game of football and all this as you see but the thing is . . . she don't wear trousers. She's just like a boy but she don't wear trousers. I'd like to be like that. She's fun.'

She was very powerful in the class, liaising between girls and boys, and teacher and pupils. Kay was mentioned by the boys as well in the same terms they would use to describe their own male friends. Here is Julie for example: 'Kay, she doesn't grass if you do something wrong. She's a good friend, she sticks with you . . . She's not stupid.' Willy likes Kay for more pragmatic reasons: '(She) supports the same football team as me . . . She's good at football . . . she's a brilliant defender . . . if the ball goes past her she'll bring you down.' Helen was more acerbic claiming that although the basis of Willy and Kay's friendship might be to do with football ' . . . she goes round with him for things she doesn't know'. Kay is mentioned by both teachers as a visible child who is strong minded and insists on what she wants from both teachers and pupils.

So Kay is much more powerful than any of the other girls so far because she can place herself in relation to the teacher, the other girls and, unusually, the boys without any loss of 'femininity' or being considered 'odd'. This brings us on to our final point which is to attempt to synthesise all the data which we have collected.

Our argument throughout has been that attainment in mathematics is much more complicated than an ability/performance model would suggest. We have suggested that attainment in itself (or lack of it) is not a unitary possession of individual children. We have shown how teachers conceive of ability or attainment in ways which relate to the ideas they have about teaching and learning as indicative of certain processes. We have shown how social relations within the classroom serve to build up positions for children. In this way children are constructed as good or poor at mathematics and this has material effects on the way in which they see themselves in relation both to their own work and the other children.

We have theorised this in terms of a nexus of positions which children fill. In this way we can see both similarities between children and the specificity of each case, not as exceptions which can only be explained as such but rather as explicable in terms of the network of positions created in the classroom. In terms of this network of positions let us look at two of the girls we have already mentioned several times.

Patricia's position, for example, in the primary school is one of invisibility. With regard to the teacher she was considered not good enough and was rebuffed when she was finally able to make contact. His view of her in turn trapped her into reliance on her friends. She was revealed as anxious about her work. The other children saw her as in need of help, which they provided whilst all the time cementing their positions as better than she. She was in a powerless position, on the periphery of the class into which she was unable to fit.

Elizabeth on the other hand filled the 'ideal' position at primary school in that she fitted squarely into the framework used by the teacher to understand good performance, i.e. her work was neat, she answered questions correctly but did not constantly need help or reassurance. In turn this gave her confidence in her own ability *vis-à-vis* both her work and her peers.

It is interesting that in the secondary school both these children replicated the positions they occupied at primary school and for the same reasons; Elizabeth fitted within the framework of relations in the classroom and Patricia did not.

Helen on the other hand was in a very different set of positions in the secondary school. At her primary school she was considered good at her work (contradicting her secondary teacher who thought she was poor) and although she was often forced, unlike Elizabeth, to seek help from the teacher and others she had few problems coping with the work. It is interesting that in her case study at primary school, a prediction was made about her transfer:

> Helen might find the transfer to secondary school more difficult because of her reliance on the definitions of others than, for example, Elizabeth who is much more self-contained.

And so it turned out. From a position of strength and relative confidence Helen moved to become more diffident, less sure of herself. She was being bullied and was no longer assured of approval from her teachers. This had left her adrift and unsure of how to slot into the positions of her class.

Fourth-year secondary

By the fourth year the children were established within their classes both in their relationship to other children and to the subject.

Throughout the school the emphasis on individual work and individual credit meant that by the fourth year the children had become totally individualistic about their work. As we showed in the last chapter, the older the children were, the less part of the school group they felt. We could surmise that working on individualised mathematics schemes for so long left them more concerned to get on than to discuss and/or to help others. Also, it seems to be the case that by the fourth year all the children were doing such different things that there was little common ground for discussion about work. Let us look at the development of the categories of analysis we have used so far.

It seems significant that the girls seen as good by the fourth-year teacher rarely sought help with their work. In this way it was possible for the teacher to see them as good, although hard working. They both helped other pupils and were seen by them as good. They are both described by the teacher as 'conscientious' even though one of them had been truanting.

For the poor girls Yong was considered to work hard but not be very good. She rarely went to the teacher for help, though, through a combination of timidity and her own feelings that she was not very good, finding mathematics a struggle. Again, as in both the classes lower down the school, it seemed to be the ability to attract the teacher's attention which differentiated the good from the poor pupils. Yong in her interview expressed her feelings as follows:

> *I'm not very good at maths . . . I haven't talked to him much about maths. I haven't asked him much . . . Sometimes when I'm not good at maths I don't ask.*

'I always struggle', she said of herself. She felt trapped, afraid to ask yet needing help. This seemed to give her a feeling of powerlessness in her mathematics lessons and she was forced to turn to others for help and support. In the video-taped lesson she spent a lot of time (nearly twenty minutes) reading and re-reading her work card. She tried to work back from the answers in the answer book. At this point the teacher came beside her:

> *T* *Let me see . . . you might, where you've got a difficult graph like this, you might find it better on proper graph paper. Do you know what I mean?*
>
> *Y* *Yeah* (The teacher moves away leaving Yong rubbing out her work) *I was all wrong.* (She tries again and attracts attention of a visitor to the classroom who goes to help) *I don't get this* (reads from work card) *You are driving 200ft behind Bruce. It says 'what is your maximum safe speed?'*

V *Right, so you just read that off the graph . . . I've not actually seen this card before. Do you read off the graph? The maximum safe speed is 55 it looks like, isn't it?*
Y *Oh yeah . . . D'you get what this means? Do you think the curve should go through the nought?*
V *Your curve does actually go through the nought . . .*
Y *Oh.*
V *Think about what nought means. If you're no distance at all behind the car in front, that means you're actually touching the car in front . . . your speed actually has to be nought for it to be safe, doesn't it?*
Y *Mmm.*

This extract expresses some of the dilemmas mentioned earlier that were experienced by both teachers and pupils. The pupil had no access to, or was afraid to express, the terms in which the activity was conducted; the teacher wished neither to push nor to do all the work for the girl, preferring to let her work out the answers for herself. This trapped both participants and, in this case, contributed to the girl's own very pessimistic view of her own ability.

Good children, regardless of sex, helped other children. Michelle asked to help a boy claimed, however: 'I've forgotten how I did it.' When the teacher came to ask what the problem was she replied: 'I can't do it anyway. I doubt if I'll be able to understand it. I don't understand any of this.' This girl was at the intersection of the constellations of being clever and helpful and seemed to want to abdicate some of the responsibility which this appeared to imply.

As noted earlier, it was most important for these pupils to be seen to be able to handle social relations properly. Helping/being helped, therefore, took on other connotations to do with being 'boring', 'unable to do it' and therefore not an interesting person. It was as 'uncool' to help as to be helped unless it was in a casual way as in this sequence between Viv and Michelle:

V *I don't get that.*
M (Who has been sitting on top of a store cupboard) *What?*
V 'x' *equals two.*
M 'x' *'ve you gotta do it on? Where? Where does it say that? . . . Is* 'x' *going along the bottom?*
V *So I put* 'a' *on two.*
M 'x' *equals two.*
V *I don't get that. First of all thought it mean a (untrans.) axis and put* 'a' *here, along the* 'x' *axis.*
M *No . . . that's wrong, that's two, that's . . . It's* 'x' *equals two, that line there.*
V *Oh God, I always get this wrong.* 'x' *equals two would be here.'*
M *Yeah.*
V *Oh I get it now, thank you.*

By this stage it seems that a lot of the children have decided mathematics is 'boring'. Few seem able to be very articulate as to why they find it so, apart from blaming themselves. We could suggest that it seems unreal. Michelle in this last extract seemed to understand her friend's problem, but makes no attempt to explain it. Equally her friend seemed content to assume that it was her fault. ('I always get this wrong.') At this stage power is invested in those who have access to mathematics as a body of knowledge and they retain power by not letting others into their secret. Those who are good have understood the rule-governed nature of the interaction and are able to work within it successfully. There was much discussion about the amount of work done and this was related to examinations since a school-based CSE Mode Three course derives 50 per cent of marks from course work for which a minimum number of matrices need to be completed. Rule challenging or following in the sense that this has been discussed earlier seemed at this level to apply more to those who challenged behavioural rules. This was not necessarily sanctioned (Michelle often truanted) but we saw little evidence of an ability to 'break set' even amongst the good children.

There was least discussion of mathematics amongst this age range. Work was done silently. Help tended to be offered practically, with little or no discussion of issues. Viv summed up a dilemma expressed by all the poor girls interviewed:

> *V* *When I came to this school I had a Triple One which you get in primary school* (refers to band one on ILEA's comparability tests given in the primary school to allocate children to ability band one, two or three, prior to entry to secondary school. This is used as a way of making sure the secondary school's intake is properly mixed in terms of ability).
> *I* *Did you enjoy maths in your primary school?*
> *V* *Absolutely detested it.*
> *I* *What happened once you got here then?*
> *V* *I don't know, at some point I thought, ugh, I'm no good at maths and I didn't really try I suppose . . . now I'm trying to work again. But also I get easily distracted by people . . . who're good at maths, anyway so they can distract us and get back on with their work again.*

However it would be unfair to suggest that the girls accepted the teacher's definitions of their ability unequivocally. As we have stressed throughout, the depiction of girls as passive is one of the areas to which we take exception in some of the work on gender-stereotyping. There were a variety of ways in which the girls in our sample resisted the teacher. Power relations within the classroom were constantly being renegotiated. Yong, for example, insisted, despite the teacher's experienced misgivings, in attending a group which was doing work more specifically aimed at the 'O' level syllabus. Often, resistance took the shape of refusing to accept the teacher's definitions of what was considered appropriate and thus refusing to ask him for help, or truanting (see Hayward 1983).

The videotapes of the boys in the sample show more interaction between them and the teacher about mathematics, with the teacher correcting work and discussing problems and answers with them in more depth. The boys were monitored more closely in relation to their work rate.

> *T* *. . . Trevor you're not doing as much as usual. Mark's obviously a bad influence.*

It is the good boys—as chosen by the teacher—who seem to have most contact with him about work. Yet, as we discovered, only one child was entered for 'O' level and that was a girl who insisted on the entry against the wishes of the teacher. We were not able to explore this in more detail because it took place after the period of our fieldwork. Like the girls, the boys discussed amongst themselves their lives outside lessons whilst doing their mathematics. All the children were doing different things so there was little common ground for discussion, although most of the children would have covered most of the areas at some time.

The teacher was understood as a facilitator and not often approached for help, although he carefully monitored the class and tried to see every child as often as possible. Yet he, too, was caught in a double bind. He worked to facilitate the children's learning by providing the resources and a suitable environment. Out of respect for their feelings and aware of the fears mathematics can generate, he tried not to push or to be authoritarian. This, in turn, left the quieter girls struggling and with feelings that they could not do the work and that the teacher was unapproachable—the very opposite of what he wanted to convey.

In part this double bind had to do with the imminence of public examinations and the necessity—since the school offered the choice—of selecting for 'O' level and CSE groupings, and thus irrevocably labelling some children as more clever.

We shall discuss in the final chapter the implication of and conclusions which can be drawn from this brief overview of our work.

Summary

In this chapter we have argued that relations of power and powerlessness in the classroom are not fixed but constantly shifting. Girls can be successful in terms of mathematical attainment, gaining power by taking responsibility in the classroom, but remain relatively powerless in terms of teachers' judgements of their performance. Since the latter depend on indications of the challenging of rules which are understood as 'real understanding', 'flair' or 'brilliance', girls are often left in an ambiguous position.

Our analysis of individual cases exemplified some positionings, occupied by a variety of girls (and boys). We focused on the dimensions of helping and being helped, suggesting that aspects of femininity, cohering around positions of being nice, kind and helpful, meant that some girls could be helpers and even the poor girls could at least succeed in being nice to others.

Some good girls have to manage a 'balancing act' between feminine and masculine positions, and the girls who are most academically successful *and* simultaneously popular with peers, achieve both. Such girls are visible in the practices of the classroom. The poor girls remain invisible, unnoticed. They do not like to ask for help (and the teachers like the girls to work out the answers for themselves). Girls can occupy positions which are related to masculinity, for example, challenging the procedural rules of mathematics. Yet, such positions are never unambiguous nor gender-neutral, because of the effects of other positionings on the girls in question.

By the fourth year of the secondary school girls are still in similar positions, although interaction with the teacher is far more restricted and mathematics has become a more individualised (and secret) activity. Many children by this time find mathematics boring.

CONCLUSIONS

1. Examining the relationship between ideas about the teaching and learning of mathematics and classroom practices produced a reading of success and failure in which actual attainment is no longer a simple or reliable indicator of success. The move towards 'real understanding' and away from 'rote memorisation' means that certain characteristics are invested in the individual. It is, therefore, possible to be successful for the wrong reasons. That is, a child may do well, but may be suspected of 'not understanding'. Teachers are,

therefore, at pains to promote such understanding in their practices and to look out and find, evaluate and remedy evidence of success for the wrong reasons. This means that their evaluations are made in terms of the presence or absence of these attributes.

Such practices have particular effects, in that the characteristics taken to be indicators of 'real understanding' are to a large extent co-terminous with those used to describe masculinity. There is, therefore, a problem for many girls: their success is double-edged; the characteristics of femininity which they display lead teachers to assume a lack of understanding. This has particular effects in terms of their classroom performance and in terms of the practical consequences of the disjunction between success and its evaluation.

2. The analysis of the data from the children's interviews, repertory grids and classroom activities suggests that girls struggle to achieve a femininity which possesses the characteristics which are the target of teachers' pejorative evaluations. We use the word 'struggle' advisedly. The data does not suggest an easy or natural process, but precisely a struggle to be understood/understand themselves in certain ways, which have particular effects on the social relations of the classroom.

Our argument is that girls are located at the nexus of a constellation of practices characterised by the tortuous and contradictory relationship between gender (masculinity/femininity) and intellectuality (academic performance and attainment). Femininity particularly appears for the girls to be related to such characteristics as 'helpful', 'nice', 'kind' and 'attractive'. These are precisely the characteristics which help to render them good and hard working in the classroom and so lead to academic success, but not to a display of those characteristics which are read as indicating 'real understanding', 'flair', 'brilliance' and so forth.

How and why the positionings in different practices cross over and work in such ways as to produce different effects in particular girls was not the object of this enquiry, but it is examined elsewhere (Walkerdine, in preparation). We believe it important to examine and produce in more detail an analysis which explores the how and why of particular individuals and their specific positioning. In other words we would argue that it is necessary but not sufficient as an explanation to understand the girls as positioned in relation to the normative and normalising effects of the practices. We still need to understand the processes by and through which this occurs.

3. We have paid particular attention to the phenomenon described as rule following/challenging. We examined the way in which

following the procedural rules of mathematics and the behavioural rules of the classroom was necessary to successful completion of tasks. However, challenging the internal rules of the mathematical discourse, relating particularly to the teacher's authority as guardian of those rules is important in producing what the teachers describe as 'real understanding'. Such challenging requires considerable confidence because it necessitates the recognition that rules are to be simultaneously followed and challenged. That many girls do not have such confidence, nor would dare to make a challenge offers a different explanation of girls' mathematical development than one which relies on the naturalistic and immutable. Such an approach as ours offers a profound challenge not only to practice but to theories of cognitive development and of masculinity and femininity.

For the full details of this study and the arguments derived from it, the reader is recommended to consult the complete Bedford Way Paper 24 (Walden and Walkerdine 1985).

REFERENCES

Bar-Tai, D. (1978) 'Attributional analysis of achievement related behaviour', *Review of Educational Research*, **48**.

Carver, R.P. (1978) 'The case against significance testing'. *Harvard Educational Review*, **48.**

Corran, G. and Walkerdine, V. (1981) 'The Practice of Reason, Vol. 1, Reading the Signs.' Mimeo, University of London Institute of Education.

Diener, C.I. and Dweck, C.S. (1978) 'An analysis of learned helplessness: Continuous changes in performance, strategies and achievement cognitions following failure.' in *Journal of Personality and Social Psychology*, **36**.

Hayward, M.S. (1983) 'Girls, resistance and schooling.' Unpublished MA dissertation, University of London Institute of Education.

Henriques, J. *et al.* (1984) *Changing the Subject: Psychology, Social Regulation and Subjectivity*. London: Methuen.

McRobbie, A. (1978) 'Jackie: an ideology of adolescent femininity', Working Papers in Cultural Studies SP 53. Birmingham: Centre for Contempory Cultural Studies.

Morrison, D.E. and Henkle, R.E. (eds), (1970) *The Significance Test Controversy*. Chicago: Aldine.

Salmon, P. (1976) 'Grid measures in child subjects', in Salter, P. (ed.), The Measurement of Intrapersonal Space by Grid Technique. London: Wiley.

Spender, D. and Sarah, E. (1980) *Learning to Lose: Sexism and Education*. London: Women's Press.

Walden, R. and Walkerdine, V. (1982) *Girls and Mathematics: The Early Years*, Bedford Way Papers 8, University of London Institute of Education.

Walden, R. and Walkerdine, V. (1985) *Girls and Mathematics: From Primary to Secondary Schooling*, Bedford Way Papers 24, University of London Institute of Education.
Walkerdine, V. (1981) 'Sex, power and pedagogy', in *Screen Education*, **38**, Spring.
Walkerdine, V. and Walden, R. (1981) 'Inferior Attainment?' *The Times Educational Supplement*, 3 July.
Willis, P. (1977) *Learning to Labour.* London: Saxon House.
Young, M.F.D. (1971) *Knowledge and Control.* London: Collier Macmillan.

PART II

What Can Be Done?

11

Should Mary Have a Little Computer?

ANITA STRAKER

The week before last I ran a workshop at a course for primary teachers who were interested in developing the use of a microcomputer within their school. Approximately one quarter of the teachers there were women, yet I know that a figure more representative of the primary teaching force as a whole would have been 75 per cent. Last week I was asked to give a talk to a group of teachers from secondary and middle schools. The teachers all had a specialist interest in craft and design technology. There were three women and twenty-four men in the group. This week I attended two meetings. One was a regional meeting of advisers and advisory teachers with a responsibility for computing within their particular local authority. The other was with a group of HMI and advisers who were planning a course on the role of the micro in mathematics. At both these meetings I was the only woman in the room.

It is not difficult to gather evidence that at present, in the field of education, more men than women appear to be attracted to or appointed to specialisms which involve microtechnology. It is far less easy to ascertain the reasons for this state of affairs, or to assess the impact which it could be having on the school population.

At the course for the secondary and middle-school teachers I asked the course members to what extent girls within their schools opted to follow design technology courses, or (where they were offered) optional courses in electronics or computer studies. In all cases the answer was the same: only rarely. When I suggested more positive discrimination in favour of girls a heated discussion broke out! One of the teachers felt that the numbers of girls on these courses merely reflected the job market, and that to encourage more girls to follow them in school would be raising false hopes of employment. Another teacher suggested that there would be little point in positive discrimination in the secondary school, since by the age of twelve or thirteen it is too late. By then girls already have a self image which, for many of them, shies away from scientific, mathematical and technological activities.

Whatever one may think of these *laissez-faire* attitudes one message at least is clear. Everyone connected with primary education, whether teacher or pupil, parent or governor, publisher or programmer, needs to be aware of the position and ready to take remedial action. We must never let it be true that by the secondary school stage 'it is too late'.

A report on *Computers in the Primary Curriculum* published by the Microelectronics Education Programme stated (paragraph 1.1.4):

> There are increasing signs that computers are being used more by boys and male teachers than by girls and female teachers. Primary schools may need to take positive steps to ensure that both sexes have equal opportunities.
>
> (Microelectronics Education Programme 1984)

The House of Lords Select Committee on Education and Training for the New Technologies suggested that:

> . . . the time when educational reform is most urgently needed is in primary school and the early years of secondary school.
>
> (1985)

This statement is borne out by a number of indications. Tony Ballerini, for example, describes a small group of five primary children constructing a model of a bell tower with a hoist, which they were planning to control from the micro.

> The girls diligently constructed the frame while the boys busied themselves with the winding engine and subsequent computer control. This was not due to a lack of interest on the part of the girls. They were bustled out of the way and were too polite to bustle back. They liked to be *invited* to use the machine. Once allowed to the fore they were as excited and as adept as the boys.
>
> (Ballerini 1985, p. 15)

What are the positive steps which could be taken by a primary teacher faced with such circumstances? Having a discussion with the children themselves could help. Supervising sufficiently to make sure that girls and boys get equal amounts of keyboard time is another possibility, with perhaps the girls taking the first turn. A third possible action is to ensure that the technological problem, in this case the construction of a bell tower with a lift, is of equal intrinsic interest to both sexes, and, if not, forming single-sex groups for the particular activity.

The same class of children had also been encouraged to bring models from home. The boys, in particular, had brought in Lego vehicles, some of which were powered by motors. It is quite possible that girls who do not have access to technical Lego at home (hinges, geared wheels, motors, and so on) might feel at a disadvantage in such circumstances, and even more inclined to think that matters technological are not for them.

The reason why boys, rather than girls, are more likely to have technical or mechanical toys at home is probably one of social tradition, and changing this is an inevitably slow process. Many parents and grandparents, educated a generation ago, have unthinkingly purchased a toy crane, a train set and a construction kit for their primary aged son, and a doll, a washing-up set and a nurse's uniform for their daughter.

One of the secondary school CDT teachers at the course described how he customarily sets for homework a technical problem which can be solved by using junk material. Nevertheless, he consistently receives letters of complaint from parents who believe that the activity is inappropriate for girls!

Interested to ascertain the extent to which the Christmas 1984 boom of computer sales had affected the number of micros children have at home, the MEP Primary Project conducted a sample survey in the early part of 1985. The 187 techers who helped us were simply some of those whom we met on courses during January and February of 1985. About one quarter of their schools were situated in the north of England, one quarter in the East Midlands, one quarter from Merseyside, and the rest from the Home Counties. There were, perhaps surprisingly, no particular differences in the numbers returned from each of the three areas.

Two age groups were surveyed: 1926 six- to seven-year-old children in top infant classes, and 2186 third year juniors aged nine to ten years. The figures are shown in Tables 11.1 and 11.2.

These surveys could, of course, have been far more extensive and far more rigorous. But I suspect the trends would have been much the same. Twice as many boys as girls will have access to a micro at home. Nearly twice as many will have a calculator. Far more boys than girls will have a digital watch. And the group of children who

Table 11.1 *Top infants survey*

	Girls from girls-only families	Girls from mixed families	Boys from mixed families	Boys from boys-only families
Owning their own calculator	18%	26%	35%	36%
Having a calculator in the family	52%	73%	71%	70%
Owning a digital watch	28%	31%	67%	64%
Having a computer in the family	19%	25%	39%	42%

Table 11.2 *Third year juniors survey*

	Girls from girls-only families	Girls from mixed families	Boys from mixed families	Boys from boys-only families
Owning their own calculator	28%	29%	54%	58%
Having a calculator in the family	66%	72%	74%	72%
Owning a digital watch	51%	55%	79%	80%
Having a computer in the family	27%	36%	48%	59%

are least likely to have any of these are girls who are either only children or whose siblings are also girls.

Familiarity not only breeds contempt. It also breeds confidence, and from confidence springs enjoyment. Children who come to school having already had experience of playing at home with wheeled vehicles, clockwork or battery powered toys, a calculator or a computer, are surely more likely to enjoy and benefit from the technological, scientific and mathematical experiences they will have in school.

Perhaps the most important of all the positive steps which a primary school should take is to conduct its own survey about the access to microtechnology which its pupils have at home, and to discuss carefully the outcomes of the survey with both children and parents.

REFERENCES

Ballerini, T. (1985) 'Ding Dong Bell' in *Primary Teaching and Micros*. March.

Microelectronics Education Programme (1984) 'Computers in the Primary Curriculum', Conference Report.

Select Committee on Science and Technology (1985) Education and Training for the New Technologies. London: HMSO.

12

*The Girls in my Tutor Group Will Not Fail at Maths . . .**

BARBARA BINNS

The results of the third-year assessment test frightened me. With few exceptions the top half were boys and the bottom half were girls. I had been worried by the lack of enthusiasm of girls and their limited participation in lessons, but had not gone beyond trying to be a little more aware of who I was spending time with in class, and concentrating on everyone rather than only the demanding few. These are mainly boys.

The information I received from the results, the realisation that this implied that few girls would even be entered for 'O' level made me see that a vague awareness and worry was not enough. Something had to be done to stop this happening. So I worked hard at dividing my time equally between girls and boys, asking girls questions and putting pressure on them to talk about maths rather than boys. I thought I was doing well. Both third-year classes were working hard, talking about their work and I was spending a lot of time sitting at girls' tables. Then I was observed, using a GIST (Girls into Science and Technology) observation schedule.

The results were interesting. Because the boys demanded my attention immediately while the girls sat patiently with their hands up, I was still spending far more time with the boys. And this was when I was being very aware of what was happening.

This made me wonder whether things had gone too far by the third year. What influences had affected behaviour in this way? How much

*This chapter first appeared in ATM Supplement 25. Reproduced with permission.

had I, and other teachers, allowed the situation to develop to this extent? Talking about this issue at the ATM Conference helped me to realise how serious it is, how many other people are thinking and talking about it from the point of view of both teacher and parent. People with girls in primary school were seriously thinking about single-sex secondary schools, and ideas of single-sex groupings within secondary schools were being put forward. I feel that these solutions do not really come to terms with the problems. There must be a way of stopping girls giving up, accepting failure and a secondary role in society.

The girls in my second-year tutor group have suddenly migrated to the tables far away from mine. They have started wearing make-up and the conversation is about boys. They have also stopped talking to me. Their maths folders are very incomplete. No girl has done the most recent topic adequately, whereas all the boys have. This has not happened before. I felt anger while reading their work and decided that I was not going to allow the girls to give up.

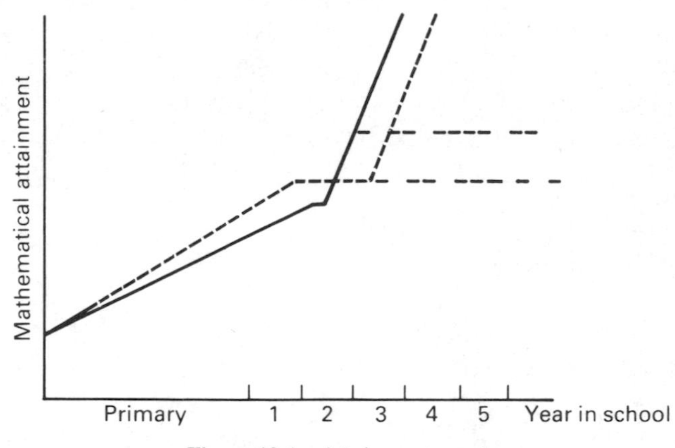

Figure 12.1 Attainment graph.

Next morning in tutor period I drew a graph something like Figure 12.1. I explained the axes and then we discussed the two lines. They did a pretty good bit of graph interpretation (well, the boys did) and after ten minutes of discussion someone asked suspiciously if the blue line was girls. We could then discuss very explicitly whether this was a fair representation of attainment. Were the girls better in primary school? Why was the dotted line flat for so long? Because girls are interested in boys, not maths, I was told by a girl, in no uncertain terms. The others all agreed. A colleague had a similar discussion with his third-year group. They said it was because girls had *problems*. They would not say what they were. When does puberty become a problem rather than an excitement?

I pushed the consequences of the graph quite hard. If they gave up now, fewer girls would pass 'O' level and 'A' level, and all the wonderful jobs that required these qualifications would be taken by men. Men would continue to rule the world while women continued to do the menial tasks. Yes, of course I got carried away with it, but they did respond. The girls were indignant at the unfairness they saw in our male dominated school and society. The boys were thoughtful about it too.

Next day I asked them to draw their own graph; as it had been and as it would be and explain why it was like that. The graphs all went up steadily in the future with one exception. Hilary's stayed flat for a little bit before rising. She explained that it would stay flat until the fifth year, then she would work again. I suggested that she might miss a lot in the remaining weeks and that James did not actually walk past the window very often. She wasn't convinced but did agree to try to concentrate on work.

Later they came into maths and within a very few minutes they had all (and I mean all) started working. Several of them had moved to different places which meant they were back in mixed groups. Good. The girls looked happy and industrious and kept asking for help. The boys were far less demanding and more relaxed. I spent every second doing maths with them.

That was a week ago. I feel that our relationship has taken a big step forward. The girls are talking and smiling and the boys are less silly. At the moment I feel it is a fight we are fighting together; they have taken their share of responsibility, so maybe the good resolutions will last longer.

I know this is only a beginning and there are many more questions to be asked. But I am determined that the girls in my tutor group will not fail at maths.

13

Experience with a Primary School Implementing an Equal Opportunity Enquiry

HAZEL TAYLOR

In the outer London Borough of Brent, an action programme for developing the conditions in primary schools in which it will be genuinely possible to claim there are equal opportunities was devised in 1983. One of the strands of that programme was the funding of an action research project in an infant school, with the purpose of devising and testing strategies for practice which would foster good, confident girl mathematicians. There was awareness of the various and not wholly consistent theories seeking to explain the general profile of girls' mathematical performance, and a concern that the common-sense view that 'girls do fine at maths in the primary school' concealed many factors which are crucial if girls are to feel they are capable mathematicians, to find the subject interesting, and to feel that it is useful. These are factors which relate to girls' perceptions rather than their observed performance and which are central to decisions they will make about their relationship to mathematics as they grow older. They are as likely to be developed through events which are part of general classroom life as they are from events taking place in the course of identifiable mathematics time, if not more so. However, the complexity of the issue is not necessarily readily recognised by teachers whc have learnt nothing about gender variables in their initial training, who perceive infant girls as more mature and conformist than infant boys, and who are generally far too busy during the day to observe and reflect on what is happening in the classroom beyond their immediate involvement. The action research project was deliberately set up in a school in which the teachers were not initially aware of the issue, and part of its purpose was to be able to involve those teachers in disseminating their experience to others who were similarly unaware.

The school was volunteered by its head teacher with the intention that the summer term of 1984 would be spent on planning the project through discussions with the teachers involved, followed by a year of classroom observation of agreed areas, decisions about strategies for change, observation of implementation, evaluation and tentative conclusions. The school received additional capitation to purchase anything that emerged as a need during the observation period, and had an extra teacher for one day a week who could be used to enable class teachers to observe and write up their observations on a rota basis. Seven teachers agreed to take part; one left at the end of the first term of observation, another left at the end of the second term. The head teacher who offered her school had herself left before the planning period started, and the work in the school began while there was an acting head who later was appointed as head teacher. The project was therefore subject to rather more than usual staff changes, and was promoted by the new head in a very demanding period for her. It could not possibly be claimed that the teachers' involvement was only possible because of ideal conditions: in addition to the staffing changes already mentioned, the project was affected by two periods of industrial action by teachers. The whole of the summer term intended for planning was subject to withdrawal of good will, and in the spring term of the research year, the action began again. The effect on the project makes the point very clearly that curriculum development work is undertaken by teachers in their own time with no formal recognition of the fact.

The planning therefore took place while the head teacher twice took the whole school for an extended assembly, thus freeing the teachers to attend two meetings with the equal opportunities adviser who had initiated the project, to talk through the possibilities. A minimum of information was provided about existing research on girls and mathematics with particular reference to the APU surveys, to differences in performance in secondary schools, and to Valerie Walkerdine and Rosie Walden's research; the main time was for teachers to explore their own ideas and perceptions about what went on in the classroom. At the first meeting, teachers discussed sex differences in the use of toys, and three observations were made: that the teacher did not know which toys girls played with, that girls and boys played equally with everything, or that girls avoided Lego and other construction toys. This range of opinions reflected the range encountered amongst other infant staffs by the adviser. There was certainly no conviction at this stage that there was a particular problem in relation to girls and maths, and gender differences that were perceived were seen as natural. The teachers knew that the focus of the project was to be on observation and then on changing classroom variables to see what effect they had, and by the second meeting they decided to concentrate on examining what went on during free play. At this point much discussion took place about observation techniques, the size of groups to observe, and the way in which certain personalities might affect and distort outcomes. A good deal of anxiety emerged about the status of the research and an expressed need for expert guidance. A box of books was provided at this meeting for teachers to do some background reading both around women's rights issues and mathematics. The model of teacher as researcher is one that is consistent with a feminist perspective. It is important that teachers are supported through to valuing their own observations and insights, and to giving them status, and also to a willingness to take responsibility for determining what to examine. That determination of the issue by teachers is also crucial to their commitment to the research process, which is demanding of time and energy, to their acceptance of the findings, and to the research having a purpose for them rather than being something done by outsiders for their own purposes. The period of anxiety has to be lived through with support which directs the teachers back to answering their own questions, and with the knowledge that as the observation stage gets under way, anxiety will be replaced by enthusiasm. The professional development of the teachers through their involvement is as much a purpose of a school-based action research project as the gathering of material to disseminate to others.

At the beginning of the project year, the teachers were only partly prepared: the task had been identified as examining free play but much remained to be resolved concerning the choice of children to observe, and the manner of recording observations. There is always a great temptation at this stage to want to attempt far too much, in terms of the number of children to observe, and the range of activities. Until research begins, it is difficult to realise what a large volume of data will be generated, and how much time will be needed to sift and analyse it before anyone can formulate insights into its meanings. The tasks selected must be kept manageable in terms of the teacher time available for that process of analysis and reflection, and discussion time with colleagues and external consultants must be built in. A simple observation schedule was constructed, children were selected, and the head teacher timetabled all the staff involved to have a one-hour period of observation once a fortnight, using the additional teacher to cover. At this point, Valerie Walkerdine came to a meeting at the school to talk about her own work, and raise questions she felt were important when looking at girls and mathematics. This was a very valuable meeting because it enabled people to examine a number of unquestioned attitudes and to put the task they were embarking on in a wider context.

The periods of general observation continued for the whole term. Several discussion meetings were held, and much informal conversation took place. The head teacher organised the keeping of records so that each teacher kept observation schedules and commentary and reflection notes after each period of observation in a form that was easily accessible for everybody. Very quickly, the one-hour observation period became forty-five minutes' observation and fifteen minutes' time for adding a commentary. A major review meeting was held at the beginning of the Spring term, to look at the findings and decide on action.

The findings and subsequent events will be written up in detail at the end of the project; in this interim paper it is only possible to indicate the general directions of further exploration that the observations suggested. There were common strands in each class, while the emphasis and detail changed from class to class. It was clear across all the classes that overall boys played for longer with Lego and other construction toys and that among a number of reasons for this was the exclusion of girls from that play by boys. Anne, the nursery teacher, noted an exchange in which two girls who were dominant in home corner play tried to join some boys in brick play by asking, 'Can I sit here?' and 'Can I get on the bus?'; they received negative answers, while an additional boy was accepted into the game. The girls' style of trying to join in is an important factor in their rejection: pausing to ask permission before sitting down or getting on the brick bus gives an ideal opening for exclusion, and at the same time follows well-documented patterns of female permission-seeking which reflects acceptance of less power than other groups. Anne felt sad and frustrated by her girls—'I wanted to shake them!'—as she observed girls who were dominant in their own group dominated by boys. She identified with their behaviour and knew how she had felt when younger. Other teachers also reported this pattern. It was also reported that when faced with an unfamiliar activity, girls who were identified as able, and were dominant in games, nevertheless came to ask the teacher what to do next. They knew what to do, but needed reassurance. Not all girls, however, fell into this category, and a second important observation was of the effect of different personalities and styles of dominance on what happened in groups of children working together. Several girls were observed across the classes who the teachers knew did not have a thorough understanding of the mathematical concepts involved in the work they were doing, yet who gave every appearance of understanding and leading. One girl was observed by Iris in her middle-infant class who was noisy and dominating in an all-girls group, and who gained information from the quieter girls and gave it out as her own. Her grasp of the maths involved was weak. When placed in a mixed pair with a quiet boy, she simply did as she was told. Several examples were also found of quiet girls who when actually asked to organise a group or put in a teacher role, did so extremely effectively. The teachers felt that personality and style of dominance were issues that needed closer investigation, in both single-sex and mixed groups. The third main observation to emerge from the first term's sessions was that when girls and boys played with Lego or other construction equipment, they tended to do so in markedly different ways. Andrea, who had a reception class, had recorded many instances of girls and boys playing with Lego. She noted that girls were reluctant in free play time to choose building unless the boys were removed, while with boys it was a popular constant choice. She also noted that more boys than girls approached the teacher with items to show, and most importantly, that while building, there was a definite pattern for boys of predicting what would happen as they built, and of talking themselves through their activity. Connected to this, there was a likelihood that the boys were building items to be part of an already devised imaginative game, so that the construction activity had a definite purpose beyond its own sake. The range of things constructed by boys was very wide, and nearly always they were things that would move. For girls, this was not so: they sat down to make an item (usually a house), talked about something quite different, and got up and moved on to something else when they had either finished or were bored. This pattern of behaviour was reported by teacher after teacher and led to speculation about both its cause and its effects. How does boys' talk while building compare with girls' talk when acting out domestic play? What central engagement with the activity, and internalisation of its meanings, takes place through and is expressed through the narrative constructed around it? To what extent does a later comfortableness with science and engineering come through the imaginative engagement with the act of making cars, helicopters, fire engines and space ships out of Lego and Meccano, rather than through the familiarity of three dimensional construction from frequent practice? What is it about the activity that brings boys back to it again and again, and has them clamouring to do it if they are temporarily excluded? Jan Harding and Martin Grant's work on girls and technology has already shown that they engage in it for different purposes from boys, girls being attracted to solving a problem in a social context and for a helpful purpose, and boys being attracted by the making activity. Harding's current work in tracing the reasons for girls' style of attachment to science back to the earliest stages of gender identity development and the ways in which girls and boys separate differently from the mother would seem possibly to be supported by the style of language/play interaction noted for the boys and Lego. If a social context enabled older girls to involve themselves in technology, could younger ones be encouraged to build more by providing play contexts for them that might seem similarly satisfying? There was obviously far more to Lego play than simply ensuring that girls and boys both did it. Decisions were therefore taken to develop strategies for encouraging girls to fit construction play into an imaginative context, while at the same time for half a term making drawing and cutting boys-only, and Lego girls-only activities. The next stage of the project was thus to implement strategies over the term with that aim, and observe what happened.

The teachers were free to decide for themselves what to do, but all developed varieties of the same basic intervention of sitting with a group of girls or a mixed group with Lego and trying to develop the context in which their building was taking place. This stage of the project will inevitably need to continue for some time, because changes do not take place rapidly, either for children or for teachers. A variety of approaches seems promising. One teacher developed a range of work with her class using the story of the three little pigs. As part of that work, she asked groups of girls to make the third pig's house, the house of bricks. Other groups in the class were making houses of straw and sticks. The girls, with this clear purpose, completed houses, which were lavishly praised by the teacher and put on display. They acquired status within the class through this activity, and saw Lego building as something which had been validated for them. Clearly the incorporation of building into work around a particular story is a way of involving girls both imaginatively and practically, and provides incentive to finish—there was much reporting from other teachers of girls beginning houses but not finishing them. Other teachers specifically asked all girl groups to make 'something that moves', which produced a variety of items—a wheelbarrow, cars, a fire engine, a truck, a helicopter, and a house with wheels, but did not lead to the incorporation of those objects into other play. Teacher presence was a crucial factor in developing opportunities for extending the activity, however, as in an example when a girl in Lynn's group made a bridge. Lynn asked what might go under a bridge, so the girl made a boat. The boat wouldn't go under the bridge, so Lynn asked what needed to happen to the bridge so that the boat could go under. The girl, with continued teacher interest, made two efforts to increase the height of the bridge until the boat could go under, and was well pleased with the results of her efforts. She asked for her work to be displayed and it was.

Another strategy was to attempt to develop group play round an imaginative theme, perhaps by providing a street layout for which to construct various buildings. One group of girls were building a hotel and their play was extended by being asked how the guests got to the hotel. This led to the building of taxis, trains and planes and to talk around the things involved. Another, mixed, group built a park, with an ice cream parlour, swings, and other amenities. Both sexes worked excitedly at the task, led by a girl with little building skill but a powerful personality. The following session, in a mixed group without this girl, boys and girls worked separately and the boys became involved in imaginative play while the girls did not. One teacher commented that she had noticed that an idea she might suggest in one session would not be immediately taken up, but might well appear subsequently, perhaps once it had sunk in as a possibility. She had found that trains had become popular for her girls to build, and that they were played with on the floor accompanied by appropriate sound effects.

Clearly it is early days in the exploration of this particular strand of inquiry into girls and building activities, and at times to the teachers it seems a very long way from mathematics. A whole range of questions remain to be answered, not least concerning how effective isolated strategies can be without the support of a general commitment to equal opportunities throughout the school. The project so far, however, has, at the very least, made the teachers vastly more aware of what is happening in their classrooms, developed their enthusiasm for taking part, and thrown up an extremely promising line of enquiry.

14

Girls and Mathematics at Oadby Beauchamp College

ALAN EALES

This chapter sets out the development of a girls and mathematics programme in a 14–19 upper school and community college in Leicester. It traces this development from an initial general concern through various specific actions concluding with a retrospective view by some students and teachers. Although no school is 'typical' nevertheless others may gain from the sharing of this school's experience.

Acknowledgement is made to the continual support and encouragement of Maureen Cruickshank, the Principal of the college, and to Gaby Weiner and Valerie Milman formerly of the Schools Council. Some parts of the text have a shared authorship, particularly with Linda Oxtoby and Veronica Warner, both from Beauchamp College, and with Margaret Clifton, now at the City of Leicester School.

MAKING A START

In June 1981 Maureen Cruickshank, the Principal of Oadby Beauchamp College, Leicester, invited Gaby Weiner to talk to an open meeting of staff. Gaby was then co-ordinator of the Schools Council Programme 3 'Reducing Sex-differentiation in Schools' and was looking for schools to work with her in the project. The meeting was attended by about a third of the staff, including a number of mathematics teachers. This meeting was open and entirely voluntary and amongst those who attended there was no disagreement about the need to tackle the problem of sex-differentiation in schools. There was, however, considerable uncertainty about what to do and how to go about it. For example the natural grouping of teachers was in subject departments but some problems needed a wider, more general, approach. In fact the teachers attending this meeting split into various groupings to consider what to do next. The mathematics teachers decided to stay in a subject group but to keep in touch with the other developments. Even at this early stage it was recognised that to be effective the strategies adopted would need to be coherent and comprehensive.

In the mathematics department the typical approach to innovation was to encourage and support those teachers who were interested and for them to make written and verbal reports at monthly department meetings. At one such meeting the teachers who had attended the initial Schools Council Project meeting declared their wish for the department to become involved and it was agreed to review the literature, sound out opinion within the department and present a paper which would include specific proposals. This approach ensured that any negative or destructive comments, which sometimes accompany the early stages of an innovation would be voiced in private and could be talked through with individuals. The programme of literature review, opinion testing and proposals was undertaken by Linda Oxtoby and the author.

Those outside the mathematics department had chosen various aspects of sex-differentiation to explore, including attitudes of local employers, mixed physical education, English and science results and the difficulties faced by boys in studying home economics and languages. Although there were cynical and even antagonistic reactions from some staff (none of them mathematicians) there was sufficient support for the schools involvement in the Schools Council project to get under way. This agreement by about twenty or so staff (out of eighty) to begin an active exploration of the situation at Beauchamp was not continued when methodology was discussed. On the one hand scientific rigour was considered desirable in order to monitor what was currently happening and to evaluate any intervention but on the other hand some teachers felt they would be unnecessarily hampered if they had first to learn (what seemed to them to be) obscure statistical and scientific methods. In fact after much agonising public debate, exhaustion set in and there was a tacit agreement that all teachers would utilise whatever strengths they had. In retrospect this debate seems like time wasted but it appeared important at the time.

A FIRST REPORT

The report was presented to the mathematics department and was to lay the foundation for the work for the next two years. It took the form of

(a) a summary of the current position;
(b) a digest of the literature; and
(c) some suggested courses of action.

The literature study was disappointing in that there was little non-American work to draw on.

Sex differences and mathematics at the upper secondary level: A first report, November 1981—Alan Eales and Linda Oxtoby

In the primary school girls do better than boys at mathematics; at the secondary stage however there is a progressive deterioration of this lead until eventually by the age of 18 the positions are reversed. This effect is often explained as merely a limit of ability but it may be that factors quite distinct from the individual's ability are having a significant effect on performance, and that these factors affect girls more than boys.

The difference in attainment between boys and girls is relatively small for CSE results. At GCE 'O' level and 'A' level examinations, boys achieve more than girls and when maths becomes optional more boys take it than girls. Fewer women than men pursue careers in maths and most of those who do, do not achieve equally with men in employment status.

There may be a biological basis for sex differences in mathematical ability but we have decided not to deal with this question directly and we are going to look more closely at social reasons. If there are biological and social reasons then the two factors may be closely linked. A young girl's view can be influenced by parents, teachers, peers, the general school organisation (and society in general) and it would seem that the general message which some girls accept is that maths is a male domain.

1. A study of numerous articles, but mainly the three publications listed below have led us to these conclusions. These three books are particularly useful. The first gives an excellent review of all but the very recent work in this field and the others a fascinating insight into alternative methodologies besides conventional controlled experiments.
2. Whereas we shall use statistical analysis wherever possible we shall accept, particularly in the early stages, subjective interpretation of small samples or even of anecdotes.

3. We shall at least set up a base for future analysis even if current analyses are limited.
4. We shall focus in particular on three areas:
   (a) Comparison of performance on any maths tests, exams, etc.
   (b) Intervention to remedy (supposed) prejudices produced by existing class grouping, setting, etc.
   (c) Longitudinal studies of individuals.

Literature

1. *Women and the Mathematical Mystique*, edited by Lynn H. Fox, Linday Brody and Dianne Tobin.
2. *The Practice of Reason*, Volume 2, 'Girls and Mathematics', Rosie Walden (Eynard) and Valerie Walkerdine.
3. *Do you Panic About Maths?* Coping with maths anxiety, by Laurie Buxton.

Some points from *Women and the Mathematical Mystique*

Parents (page 197)
1. Parents often have lower educational expectations for daughters than sons.
2. Parents have a greater acceptance of low levels of achievement in maths for girls and boys.
3. Parents re-inforce sex-role stereotypes in their choice of toys.
4. Parents are more likely to think of maths as a more appropriate career for boys than girls.
5. Help for maths homework is more frequently sought from their father than mother.
6. Parental or family support is remarked upon in the background of women mathematicians.

Teachers (page 197)
1. Teacher encouragement is claimed to have been important to many female mathematicians.
2. Many teachers see maths as a masculine domain.
3. The teachers' sex seems to be less important than behaviour and attitude.

Pupils (pages 197, 198)
1. More American high-school students believe boys are better than girls at maths.
2. More American high-school boys than girls are likely to agree with 'Maths is male domain'.
3. A female peer-group support is necessary to achievement and course-taking.
4. A critical mass of girls might be necessary to encourage entry to advanced courses in maths.
5. The early identification and 'tracking' of the mathematically gifted may be crucial to their later willingness to continue studying maths.

6. There may be a need for occasional groupings of girls if only to explore feelings and re-assure girls that they are not unfeminine or odd because they like maths.
7. Students need better counselling and career education.

Some points from *The Practice of Reason*

Pupils (pages 28, 3, 70, 76)

1. The fit between types of activity used to teach maths in primary school and the kinds of roles held to be suitable for the sexes is much closer in primary school than in secondary school so there is little contradiction for girls between seeing themselves as female and being good at maths.
2. Up to and including the age of 11, girls do slightly better than boys on tests of verbal reasoning, English and number work. Boys do slightly better on non-verbal tasks.
3. Girls confident at maths are also confident in other subjects and are seen as being good at their work by their peers and teacher.
4. Mathematically successful children are active, assertive and confident.

Teachers (pages 1, 5, 64, 83)

1. Infant school teachers more easily allocate girls to the two extremes of mathematical performance than boys.
2. Very often teachers explain girls' good performance as just rule-following.
3. Teachers favour children who seem to understand the task set and are able to do it correctly.
4. Infant teachers give most attention to pupils who are confident—regardless of sex—but most of these confident children are girls.

Some points from *Do You Panic About Maths*

Pupils (pages 124, 125)

1. Girls are generally more protected than boys and are expected to be more obedient.
2. Much of the work pupils do in maths is either right or wrong, so the maths teacher might represent more of an authority figure than other teachers. This might induce anxiety in (obedient) girls.
3. If girls are brought up to obey rather than to challenge rules it might mean that they are well prepared for maths in the primary school but the attitude of these girls might hinder them later on.
4. In the secondary school, some girls might see maths as a subject that makes too many demands and needs too much exploration at an age when their general mood is one of caution.

There was considerable discussion over a basic methodological dilemma. It was desirable to use good experimental design and practice so that it was clear what effect was generated by any action taken. On the other hand controlling variables in educational research has an intrinsic snag; a tightly controlled situation is artificial, rendering results meaningless, but a lack of control makes interpretation of results virtually impossible. Reluctantly it was accepted that a compromise of monitoring variables and making limited subjective judgements was the best that could be done. In this light three main activities would begin.

1. Analyse performance of Beauchamp students in tests, exams, etc.
2. Intervene to remedy prejudice.
3. Make longitudinal studies of individuals.

(The first of these was necessary in order to have a base line for future comparison.)

STATISTICAL WORK

It was vital to know in some detail how the performance of boys and girls varied and a considerable amount of time, particularly in the first year, was spent in analysing results. Analysis of results in the future would be less time-consuming once a start had been made. (Where possible sixth-form girls were involved in making the analyses so as to use their skills while simultaneously heightening their awareness.) Even within the department there was a statistics phobia parallel to the mathematics phobia amongst others; so notes on statistical tests were produced particularly on the setting up of hypotheses and one particularly useful, but simple, test.

The general research hypothesis was:

> that boys and girls perform differently in tests and exams

and so the general null hypothesis was:

> there is no difference in performance between boys and girls in tests and exams.

Some examples of specific null hypotheses were:

> That scores of boys and girls in a general test taken early in the fourth year would not be different.
> That 'O' level grades A/B/C *or* CSE grade 1 (after allowing for dual entry) would not differ with sex.
> That mathematics 'A' level results for Beauchamp girls (boys) followed the national pattern.
> That the other 'A' level grades of mathematics students would not vary with sex.
> That pass/fail results for mathematics 'A' level would not differ by sex.

These are quoted to give an idea of what can be tested. It is important that each school makes tests of particular relevance to them and at Beauchamp well over twenty tests of various types were made. Among the significant results found at Beauchamp were:

November 1981 fourth-year test (for 213 relatively able students out of a population of just over 400).
5 per cent significant, boys better
Summer 1981 'O' level A/B/C/ *or* CSE 1 after allowance for dual-entries (on an entry population of just over 400).
2 per cent significant, boys better
Summer 1981 'A' level pass/fail (on an entry population of 93)
1 per cent significant, boys better

A FREAK RESULT

In retrospect the results of such analyses can seem trivial. 'I knew that all along. Why didn't you ask me?' says the experienced member of staff. It is possible, however, that previously un-noticed phenomena crop up and a case in point arose with the 'A' level results. The national figures (for 1978) Pure and Applied (combined) 'A' level results were compared with the Beauchamp SMP 'A' level results (for 1981) separately for boys and for girls. The 'expected frequencies' were calculated using the national proportions. (See Table 14.1.)

Table 14.1

Boys				Girls			
Grade	A/B	C/D/E	F/O	Grade	A/B	C/D/E	F/O
Beauchamp	25	25	15	Beauchamp	8	5	15
Expected	17	27.5	20.5	Expected	6.2	13.3	8.5

Even without applying any tests the low C/D/E section for the girls stands out, as does the high number of failures. A chi-squared test gave an 8 per cent significance for the boys and a 1 per cent significance for the girls. These also appeared to be in opposite directions for the lower grades. Regardless of whether the statistics have an appeal the marked bi-modal distribution for the girls warranted further attention. A file was drawn up showing, for each 'A' level mathematics student:

(a) maths and science 'O' level/CSE grades;
(b) other 'O' level/CSE grades;
(c) maths and other 'A' level grades;

(d) career intentions (particularly HE);
(e) staff comments on the 'A' level grade.

This showed how difficult it is to generalise as for each student there were particular individual reasons for low grades. The exercise did however prompt two themes. One was that two years before when these students had been interviewed on entry into the sixth form there had been an emphasis on choosing subjects because of enjoyment. For mathematics this brought a higher than average number of girls with 'O' level grade C or CSE grade 1. Subsequently, however, these weaker 'A' level students were not given any special help and they experienced increasing difficulty with homework and their test results deteriorated. Unfortunately, this was not fully appreciated by their teachers at the time perhaps precisely because the students were still enjoying their mathematics. The lesson to be learned is that any intervention, particularly where choice of subjects is concerned, must be followed up and continued support given.

A discussion about the excellent girl students elicited the comment from a physics teacher that there were always excellent girl students at 'A' level but again the middle and lower grades were less well filled. This is possibly because students taking three sciences (including mathematics) receive overlapping reinforcement of mathematics and physics skills and that only very determined girls opt for three sciences. These same determined, and very able girls, are also likely to be more confident in their own abilities and be less susceptible to peer-group and social pressures.

TWO QUESTIONNAIRES

There was an increasing realisation that as yet the students were not themselves being reached and so class discussion was encouraged. Questionnaires were used not only to gain information but as starting points for these and individual class discussions. (See Tables 14.2 and 14.3)

Table 14.2

Your maths teacher's name .... Your name .... m/f
Tutor group ....
For each question please tick the column that fits best how you feel.

	Strongly agree	Agree	Undecided	Disagree	Strongly disagree
1 Teachers give more attention to boys than girls.					

Table 14.2 (continued)

		Strongly agree	Agree	Undecided	Disagree	Strongly disagree
2	My maths teacher is stricter with boys than girls.					
3	My maths teacher gives more help to boys than girls.					
4	My maths teacher gives more encouragement to boys than girls.					
5	My maths teacher expects more of boys than girls.					
6	I think boys are better behaved than girls in class.					
7	I would learn maths better in a single sex group.					
8	I would rather be taught maths by a teacher of the same sex.					
9	In class, I like to work with students of the same sex.					
10	My family think that maths is more important for boys than girls.					
11	I think a career is more important for boys than girls.					
12	I think that girls who are good at maths seem unfeminine.					
13	The students who are best at maths in my class are boys.					
14	When it comes to doing a maths problem I get all the formulae mixed up.					
15	When I do well on a maths test I consider myself lucky					
16	I can do the work in class but I don't know how to apply it					
17	Maths is easy for me					
18	A lot of topics we study in maths makes no sense to me					

Table 14.2 (continued)

	Strongly agree	Agree	Undecided	Disagree	Strongly disagree
19 If you're a careful worker you'll find maths easy enough					
20 You won't be able to get on in life without a good knowledge of maths					
21 I don't see the value of most of the maths we do					
22 When you're thinking of a career maths is more important for boys than girls					

Each teacher then entered the results into a grid which highlighted any differences between boys and girls.

This grid was on a sheet of A4 and for each question on the questionnaire there was a space to draw up a tally chart of the responses. For one of the author's classes, for example, these grids occurred.

11. I think a career is more important for boys than girls

Boys	Girls	
///	/	Strongly agree
/ HHT		Agree
//		Undecided
//	////	Disagree
////	HHT //	Strongly disagree

13. The students who are best at maths in my class are boys.

Boys	Girls	
/	/	Strongly agree
HHT HHT	///	Agree
HHT	/	Undecided
/	////	Disagree
	//	Strongly disagree

17. Maths is easy for me.

Boys	Girls	
		Strongly agree
// HHT HHT	/	Agree
///	///	Undecided
//	HHT /	Disagree
	/	Strongly disagree

There was no need to print either the questions or the response categories in full, so a complete set of class results were recorded on a single A4 sheet. On occasion the assimilation of results from questionnaires is difficult but this recording device enabled the teacher to learn very easily from the results. This particular class was intrigued by its attitude, and a very useful discussion developed. This discussion highlighted for the boys as well as the girls how their attitudes might influence class behaviour, and what support the students gave to each other.

Table 14.3 *Example Two*

Mathematics Self-assessment Form. Name: ....(M/F)

Tutor Group: ....

Maths Teacher: ....

COMMENT	YES	NO
This term I have found the topics quite easy to understand		
Recently I have worked quite hard in maths lessons		
I have achieved some good test results for me		
I have found it difficult to work at home		
If I had taken more trouble with my homework I would have done better in tests		
I have tried hard to get things right		
I have asked my teacher for help		
When I get stuck I find it helpful to ask my friends		
On the whole I have enjoyed maths this term		

Topic	Enjoy	Do not enjoy	Understand	Do not understand

The reverse of the form was blank except for:

> Now write a few sentences to describe how you got on with maths this term and mention how you might make improvements.

On the face of it this does not seem like a form exploring 'equal opportunities' but in fact it was extremely useful in drawing out how each individual felt. It is not sufficient to profess an interest in the views of the students without giving them the opportunity to express their views to you, and as always, good teaching practices will tend to reduce sex-dependent effects. The freestyle comments called for on the back of the form highlighted a surprise effect. It appeared that the privacy of writing was valued and enabled the students to express feelings that do not normally surface. Examples were:

> *I like to completely finish a topic, finishing all the questions set and often I have found that this has taken extra work at home, as many of my class-mates work much faster.* (girl)

> *. . . but some topics I did not understand and it would not sink in at all, and that's when I give up and dread each lesson.* (girl)

> *My test results aren't bad and any bad marks are because of careless mistakes.* (boy)

> *I have understood most of the topics, I may be able to improve by working a little harder. In some tests that I messed up I feel I have understood the work but I have let myself down by making foolish mistakes through lack of concentration.* (boy)

> *When I get a good test result it builds up my cofidence and for the next few lessons I really try and work hard.* (girl)

MEANWHILE . . .

The review of the literature and the statistical results had naturally prompted discussion which, in turn, had affected the level of awareness of the mathematics teachers. Typical of this was the realisation of how boys quite naturally manipulated the classroom environment to suit their own needs whereas the girls were more acquiescent. Any teachers who felt that they were managing to give equal attention to boys and girls were urged to allow a guest in their room to observe and record their behaviour. Another teacher, a student on teaching practice or a sixth-former was recruited to be the extra person needed. It was usually found that the boys still received more attention but perhaps it had been reduced. The teachers had the illusion though that they were giving the girls more attention than the boys! Some video-recordings of lessons were made and there was also a very successful seminar led by Rosie Walden. As is often the case an outside speaker or discussion leader can be extremely useful in promoting what becomes a freer, more open debate than would otherwise occur.

It was important to cash in on any circumstances that could be turned to advantage. For example, a geography teacher wanted an afternoon to take students out of school and by using convenient 'free' periods an arrangement was made that the girls would spend one afternoon in the geography room doing mathematics whilst the boys were out and vice-versa. For this session some Fischer Technic apparatus was borrowed from the LEA and the students made moving machines. The half classes, the practical exercise and the focus on the sex-split all contributed to useful relaxed discussion of sexism within mathematics and science.

The longitudinal studies did not proceed in a formal way but at one department meeting the teachers compared their own backgrounds and a few senior students were gently asked about their family background. The department never warmed to the task of searching worksheets, text-books and examination papers for sexist references but a conscious attempt was made to ensure that all newly introduced material was not just non-sexist but used as an opportunity to encourage girls. For example, if a league table was used it could be for hockey or tennis rather than soccer. The sex of characters in questions could equally well be female as male. Items to be measured need not always be of metal, but could be of paper or cloth. Practical experience was encouraged. In 'problems' girls could buy motor-bikes and boys could buy food.

In day-to-day administration class-lists became alphabetical; public display and discussion of test marks was further reduced; competitiveness was reduced by placing more emphasis on the students' personal mathematical development. All of this could be summed up by what has now become known as 'girl-friendly' teaching.

SOME SPECIAL EVENTS

During 1982 and 1983 there was the opportunity to take part in some one-off events. Mathematics teachers visited the Open University for the 'Girls into Maths Can Go' conference, and teachers and students attended the Open University 'Be a Sum Body' conference. (A report on similar conferences appears as Chapter 15.) At this latter course the opportunity was taken for two senior girl students to run a seminar. This emphasised the importance of involving students, particularly girl students, in a positive way. Instead of talking 'at' the girls, a deliberate effort was made to involve them and to demonstrate that the school valued their skills (other than making coffee at parents' evenings!). This is of course equally true of boys but their externalising of pressures protects them, where this approach reinforced the girls' growing confidence.

There were also trips to Leicester Polytechnic computer laboratory and girls-only training sessions in the school laboratory. Both of these succeeded in that the girls enjoyed them, but failed in that there was no change in the numbers of girls continuing their involvement in computer work in school.

There were also drama group presentations during assemblies, poster displays and later a 'Girls' Day' as part of the programme for the college as a whole. In fact at the end of the summer term 1982 the author had been asked to co-ordinate this work and, in recognition of the extra work beyond being head of mathematics, was given a senior teacher scale. In this way the senior management of the college encouraged equal opportunities in a tangible way.

SINGLE-SEX GROUPS

Having established that performance of girls relative to boys in maths tests/exams progressively deteriorated at Beauchamp it was decided to construct an experimental design which would test the effectiveness of single-sex grouping in arresting this decline. The 1982 intake of fourth years was used and, partly for administrative reasons, it was decided to concentrate on the more able children. The fourth-year of 400 students was divided into two divisions, M and S. The ability of each division was similar, some allocations being made alphabetically, but there were occasions when subject choice determined the division a student was in, e.g. three separate sciences were only available in division S. It will be seen from the experimental design, however, that this difference was not relevant, and in any case was monitored.

In each division the top 90 students were selected using high school recommendations. These were split into three sets, each seeded to have a similar range of ability but one set all girls, one all boys and one, as a control group, of mixed sex. The sex, *per se*, of the teachers was not considered relevant (previous research suggested that it is attitudes which are far more critical) and for other reasons it was convenient for all these teachers to be female. The same three teachers were available to teach the parallel class in the other division but this time with a different type of set.

With the letters X, Y and Z representing the teachers this is summarised as:

Table 14.4

Division M	Division S	
X	Y	Girls
Y	Z	Boys
Z	X	Mixed

It was thought that when the allocation of students to sets was announced there might be adverse reaction and the possibility of sending a letter to parents was considered. This was rejected but it was decided that the rationale should be explained to the students and that a senior member of staff who understood, and had a belief in the work, should give this explanation. The 90 students of one division were gathered together and it was explained that certain research had suggested that girls were doing less well than boys in mathematics and that this might be remedied by using single-sex groups. The boys on the other hand were not found to do any worse in single-sex groups. (The logical comment that the girls in the mixed group might be disadvantaged was ignored and no student mentioned it!) Students were told that although they couldn't change groups there and then, they could, after a week or so, ask their maths teacher and that 'I'm sure everything will be sorted out'. This exercise was repeated with the other division.

It was impossible to ensure that the staff behaved in exactly the same way with each group, but even if the same teacher taught all the groups no guarantee of each lesson being given with equal enthusiasm was possible. It was, however, decided to teach the same topics in the same order using the same materials and that the teachers would have frequent discussions about the course. There were four mathematics lessons of 50 minutes in each week. The staff were agreeably surprised that the groups settled so quickly.

Initially a few students asked to be moved to a particular group but after being asked to 'wait on' they changed their minds. For example, one girl asked to be moved from an all-girl group to a mixed group because she liked competition from boys but she soon decided that she was happy where she was. A very able boy asked to be moved from a mixed group to a boys' group because he wanted to be with some particular other able boys. He was stalled because we were concerned to preserve the seeding of the groups. He became extremely anxious to move and stressed that his parents would contact the school. During the first few lessons he seemed agitated but he gradually relaxed and changed his mind, saying that he now wanted to stay where he was! Only seven students were moved and these to ensure they had some friends from their old school with them (this is normal practice wherever possible).

No tests were given on arrival at Beauchamp since it was recognised that settling in was important; also some students come from schools other than the feeder high schools which makes the setting of fair tests impossible.

Two of the teachers wrote up the reactions of their students.

Division M Girls—Linda Oxtoby

Several girls said that they felt more confident in an all-girl class and that it was easier to ask questions—boys would make fun. Some thought they did more work in the all-girl class than they would in a mixed class but others thought they did the same amount of work. A few girls said quite forcefully 'its great to have a break from the boys—they're very immature' and many of the girls made comments about boys distracting girls and boys not settling down to work properly. Some girls said that in an all-girl class they could have a laugh about things which boys would not find amusing.

All the girls said that they were happy to be in an all-girl class.

Division S Mixed—Linda Oxtoby

The girls said that they liked the atmosphere of a mixed class, although a few did say that they think boys tend to misbehave, and be immature and distract them at times. They said that they liked to work with friends. (It has happened naturally that all the girls work in all-girl groups except for three girls who work in a group with two boys.) Most girls seemed to think that they would not like the atmosphere of an all-girl class.

A few boys said that they did not think that being in an all-boy class would make much difference since the girls did not distract them; however, some boys said that they would definitely not like to be in an all-boy class.

All the students in this class said that they were happy to be in a mixed class.

Division S Boys—Veronica Warner

About half the class thought they were subject to more criticism by other students in the class than if girls had been present. Girls were thought to ensure a calmer atmosphere and quieter working conditions and the comment was made that girls are more sane. On the other hand some boys thought that girls tend to chat more and thus distract others from their work. Many members of the class were not particularly keen to change the present situation, observing that it would not make much difference as boys and girls tend to work in separate groups anyway. One boy commented that he had been against the idea of single-sex classes at the start but now he was pleased to be in an all-boy class. Most of the boys thought they were working harder than they would have done in a mixed class.

Division M Mixed—Veronica Warner

Most students said that they liked working in a mixed class and could see no definite advantage in single-sex classes. Several girls did bring up the question of possible disadvantage to some students if it turned out that their particular grouping was not so conducive to successful maths learning. They were reassured that if very definite disadvantages became obvious the teachers would take steps to counteract it. Many students thought that success in maths depended more on the quality of teaching than on any other factor.

One Teacher's View—Margaret Clifton

Girls—Positive

1. Most girls benefitted—the more difficult they found maths, the more they benefitted.
2. Most positive aspect was the girls' confidence in asking questions.
3. No distraction from boys, e.g. teasing about lack of understanding.
4. No class time wasted on discipline.
5. Some girls expressed the wish to have single-sex physics classes.

Girls—Negative

1. The three girls in the class who, in my opinion, have good mathematical ability, expressed resentment at the thought of staff thinking they needed single-sex classes.
2. The same girls said they missed the competition of the brighter boys.

Boys—Positive

No distraction from girls.

Boys—Negative

I found it difficult to cope with the noise of 31 boys—although it was mainly the noise of enthusiasm. I would have liked to be able to remove the disruptive element in the class as the majority were anxious to work and resented the time spent on discipline. I think this small core of boys would have been better with a few girls around!

Results

At the end of the fourth year various circumstances, particularly change of staffing, brought the single-sex groups to an end, there being a return to ability setting: one top set and two parallel sets in each division. Nevertheless an analysis of the 'O' level and CSE results of these students was made following their examinations in Summer 1984. Also the take up of 'A' level by these girls was analysed.

Introductory Test (Autumn 1982)

About six weeks after entering the upper school the students were given a test on various mathematical topics, most of which had been revised in that six weeks.

On this test there was no significant difference in the scores of girls and boys (92 girls, 86 boys using Mann-Whitney U test).

There had been a 5 per cent significant difference at the same stage in the previous year so it is possible that the single-sex groupings were already having some effect.

16+ Results (Summer 1984)

For 'O' level (grade C and better) and CSE (grade 1) after allowance for dual-entry there were no significant difference in scores between girls and boys whether comparing the single-sex groups with each other or with the mixed groups.

Table 14.5

	Single-sex groups		Mixed groups		
	Girls	Boys	Girls	Boys	
'Pass'	30	30	14	15	'Pass'
'Fail'	27	23	15	11	'Fail'
Total	Girls	Boys			
'Pass'	44	45			
'Fail'	42	34			

Note: The drop in numbers compared with the earlier test is due to transfer to other schools, illness during the exams, etc.

At the same stage in earlier years there had been a significant difference (between 2 per cent and 5 per cent) in performance of the girls and boys.

Sixth-Form Mathematics as 'A' level (September 1984)

The take-up of mathematics at 'A' level by this group was 35 girls and 50 boys. When this was compared (using χ^2) with an expected frequency based on the five previous years there was a 1 per cent significant change: a significantly greater number of girls opting for 'A' level mathematics.

Conclusion

As is always the case it is impossible to make causal links based on probabalistic statistics, but in general terms there was an improvement in the performance of the girls. It is not possible to say that this was due to the single-sex grouping, rather than, for example, all the other equal-opportunities work in the college. The change in take-up pattern for 'A' level was very encouraging. It was recognised that a continuing high level of support must be given to these girls so that previous experience was not repeated.

However, the expectation remains that the pattern of take-up is unlikely to be maintained unless a high level of support is offered to girls and the emphasis on equal opportunity is not lost.

REFLECTIONS

Two and a half years after the introduction of the single-sex groups it is interesting to note the reflections given in hindsight by both students and teachers. Five of the students, now in their lower sixth year were interviewed and asked to think back to how they felt at the time and how they feel now about the single-sex setting. These three girls and two boys claimed that the issue of sex-differentiation was not as central for them as the presence of their immediate friends in their groups and the challenge and support of students of similar ability.

On remembering the original allocation into groups most did not even realise they were being put into single-sex groups. This was despite their having been gathered together to have the project explained. Apparently this event was mostly a blur and the students simply did not take in what was being said. They were able to listen only for their own name together with checking that their closest friends were in the same group. Comments were:

> *I was incredibly worried that I would be with people I didn't know.* (girl)
>
> *I did not notice it was an all girls group until we were in a classroom.* (girl)
>
> *I just tried to pick out my name.* (girl)
>
> *I listened to where I was, and because I'm at the beginning I can soon switch off.* (boy)
>
> *You just sit down with your mates.* (boys)
>
> *I'm sure we were told that we were in single-sex groups because girls didn't do so well at maths, but it seemed to create the problem it was trying to solve.* (girl)

It seems that they did not tell their parents of the setting and indeed the school never received any complaints or comments about the single-sex grouping from parents.

The behaviour in the groups seemed to be singular to that group and teacher.

> *I thought my all girls group was much sillier.* (girl)
>
> *Different because girls seem to talk among themselves but boys mess around—a more destructive influence.* (girl)
>
> *Girls tend to sit in little groups and talk—the boys parade up and down and generally make a nuisance of themselves.* (girl)
>
> *I thought it would be noisy but in fact it was normal.* (boy)
>
> *Maths lessons are so different anyway you just can't compare with other lesson.* (girl)

There was generally a feeling of relief when the single-sex groupings were ended.

> *Pleased to be in a top set.* (boy and girl)
>
> *The whole group would work at nearer the same speed.* (girl)
>
> *I'd rather be grouped by ability than anything else.* (boy)

The questionnaires, discussions etc. which took place were seen as irrelevant or even tiresome.

> *We had to answer something like Yes/No/Maybe but I wanted to write an essay!* (boy)
>
> *They were a waste of time.* (girl)

They were depressing. I was doing alright until I thought about it. (girl)

It was difficult for the students to judge whether the single-sex groups had been of value.

We have nothing to compare with. If we'd have been in a mixed maths group (for that year) we don't know what we'd have been like. (girl)

Before Beauchamp I had no intention of doing Maths A-level. I'm not sure when or why I changed. (girl)

I'm not sure what it achieved in the end but it seemed good we should at least try. (girl)

If a girl has ability it will come through anyway in the end. (girl)

These student comments were very different from what had been expected, but, on reflection, perhaps should not be so surprising. Being in a class with your friends is more important than whether the class is all-girls, all-boys or mixed. Able students particularly are concerned with the stimulation of other bright students in their group. Questionnaires can be seen as an interruption of learning.

The teacher reflections were consistent with the comments they made at the time of the project. One noticed that after the mixing of the groups the girls seemed more self-confident and enthusiastic about mathematics. (This was borne out later by the increased 'A' level take-up.) On the other hand, by half way through the fifth year the boys had lost their originally neutral opinions on girls' ability at mathematics and replaced them by the usual lack of respect for the girls as mathematicians. The girls on the whole in their new mixed group made it very clear to the teacher that they were indignant at the suggestion that they could be put off by the attitudes or comments of the boys. It is paradoxical that this could indicate either success or failure of the single-sex groupings: success in that the girls were positive in their confidence and assertions, and failure in that they still did not understand how their education was in fact being influenced by the boys.

One girls' group was remembered as having markedly few disruptive incidents, giving more time for learning. These girls also gained in confidence both in their own skills and in being able to ask questions in class. Although realising they were benefitting, some of these girls felt they were missing the 'atmosphere' of having boys in the classroom. One boys' group was still remembered as one of the most difficult to handle, although this may have been due to a particular small group of students. A girls' group was recalled as being very communicative and constructive and perhaps this had been particularly helpful to those girls about the 20–30 percentile down, around 'O' level C/D grade. There were also fewer girls who switched off from mathematics. The boys in the disruptive group probably suffered because of the amount of teacher time spent on control.

The judgement on whether the single-sex groups were worthwhile seemed to hinge on attitudes when a 'normal' mixed situation re-occurred. One teacher felt that if the girls in this situation could ask more questions and stay confident then this was the most appropriate measure of success. This was echoed in a different way by one girl student who thought the girls' group would have failed if 'You lose the ability even to count as soon as you are in male company'.

POSTSCRIPT

Looking back over the first years of what is now a well established girls and mathematics programme there are a number of pointers towards general principles.

1. Initial wide discussion is necessary before action can be taken.
   This discussion may centre on a specific area (methodology in our case) but it is necessary to work through the preconceptions and prejudices of those involved. A common frame of reference is necessary and the particular situation must be put in perspective.
2. The support of the head teacher is vital.
   It is important that the senior management of the school gives and shows its support in a tangible way. The appointment of a co-ordinator of equal opportunities work is one example of this.
3. An overall school programme is desirable.
   The work within the mathematics department will have its effect heightened if it is part of an integrated school (or even LEA) programme.
4. The department should make its concern public.
   Although not designed for this purpose the single-sex grouping exercise was a demonstration that here was an issue of importance to the school.
5. The students, both boys and girls, must be actively involved.
   The students themselves are the most powerful resources to achieve the changes sought, and their active involvement is crucial.

6. Small organisational changes can be surprisingly effective.
   The hidden curriculum implied by such features as boys and girls lining up separately, of boys and girls being listed separately in registers, etc. can be changed.
7. Moderate, co-ordinated efforts are more likely to be successful than extreme, fragmented ones.
   Although it may be necessary to start with a group of just a few teachers it is important to draw in as many staff as possible.
8. Good practice is girl-friendly.
   This is not the tautology it seems, or at least it has not been recognised yet! Emphasis on the learner, due regard for the individual, respect for each personality, the reduction of competition but encouragement of personal achievement will all improve the lot of the girls.
9. Public comparisons of the students should be eliminated.
   Once the regard for the individual learner is established it is difficult to understand the need for comparisons, public or otherwise. However, if the school systems demand the ranking of students, even if only in sets, then the publishing of lists etc. should be minimised.
10. The separation of the sexes can be a useful device.
    The single-sex groups were an extreme case but any separation can be useful. Girls *are* sometimes inhibited in the presence of boys; or perhaps, more often, the girls choose not to demand the same attention. The separation of the sexes can make counselling of both groups easier.
11. Try not to work with just some of the girls.
    It is a fair criticism of the programme described here that it reached mainly the more able students.

In conclusion:

1. Be clear what you are trying to achieve.
2. Be clear how you intend to achieve it.
3. Evaluate, or at least monitor, what happens.

REFERENCES

Burton, L. (1981) *Do You Panic About Maths?* London: Heinemann Educational

Fox, L.H., Brody, L. and Tobin, D. (eds) (1981) *Women and the Mathematical Mystique.* Baltimore and London: John Hopkins University Press.

Walden, R. and Walkerdine, V. (1982) *The Practice of Reason*, Volume 2, 'Girls and mathematics'. London: University of London Institute of Education.

15

Girl-Friendly Mathematics

LEONE BURTON and RUTH TOWNSEND

This chapter contains a report of an initiative to change the perspective which girls have of mathematics and its relevance to their future.

The format was a one-day conference: 'Be a Sumbody'. The first two such conferences to be run in the UK took place at the Open University in the summer of 1982. The responses of the girls, and of the teachers who accompanied them, were such that the organisers were convinced of the validity of running such events. One of the present authors (LB) then moved from the Open University to take up a post as Head of Mathematics at Avery Hill College and determined to pursue the initiative. It was decided to make the event one of joint sponsorship between the college and GAMMA (Girls and Mathematics Association—to join, contact Marion Kimberley, Goldsmiths College, London) partly because this opened the network of possible contributors to the conference and partly because it would bring GAMMA to the notice of many people who might be interested in its activities.

The impetus sprang originally from a combination of factors. First, there were research studies which were suggesting that adolescent girls were more comfortable learning in single-sex classes. When offered a choice, they preferred a collaborative atmosphere in which

*This chapter first appeared in *Mathematics Teaching*, III, 1985. Reproduced with permission.

they could explore ideas without fear of ridicule, put-down, or even simply the pressure of their social relationships with the boys.

In addition, increasing attention has been paid recently to the disproportionate rate of achievement in mathematics between boys and girls in the later years of the secondary school but, more importantly, to the increasing negativism of girls' attitudes to mathematics as demonstrated, for example, in the APU findings. Finally, girls appear to be unaware of the role of mathematics as a critical filter to careers. No longer does an 'O' level in mathematics open the way only to jobs in science and technology. It is also a requirement for jobs in, amongst other things, advertising, health-service administration, personnel work, travel, and, of course, teaching, all of which might be considered areas favoured by women.

Concern with an investigational approach to the teaching of mathematics was finally also given the Cockcroft seal of approval in *Mathematics Counts* (1982). Investigations and problem-solving were itemised as two of the six necessary teaching styles to ensure a full range of mathematical experience. Interestingly, there has been no research which has looked specifically at gender differences either in performance or in attitude to mathematics explored in a wider way. However, those interested in this field report that girls and boys do appear to participate on a much more equal basis when they are in control of their own learning. And girls themselves are very positive about such experiences. It seemed, therefore, that to run a day-conference in which girls would be given the opportunity of meeting mathematics from a different perspective to that normally found in schools and in which the mode of working would support and encourage them might help to start a new perception of the subject and the individual's relationship to it. Putting this together with some information on careers seemed like a good way of motivating girls to re-think their position with respect to mathematics and, perhaps, to its relevance to them.

The day was organised, therefore, from a wide range of concerns. We were concerned that girls achieve less and feel worse about mathematics as a subject of study than boys. We were concerned that, in a time of rising unemployment, failure to achieve an 'O' level in mathematics puts girls at a distinct disadvantage in the jobs market. We were concerned that mathematics as encountered at school gives girls, and boys, a skewed view of the subject and their relationship to it and, finally, we were concerned that the absence of a problem-solving/investigational approach might disadvantage girls, as well as boys. In the climate of an active initiative on equal opportunities in the ILEA, we were confident that schools, and pupils, would be interested in ways of implementing a more egalitarian approach to mathematics education. In the event, we were nearly killed in the rush!

ORGANISATION

The first 'Be a Sumbody Conference' jointly hosted by Avery Hill College and GAMMA took place in March (1985). The audience was girls aged 12–14—it was felt that these are the years when the decision to disengage from mathematics is made. A first 'flyer' saying little more than this was sent out to secondary schools in the local authorities near the college asking teachers to telephone and book places. Within two days we had been persuaded to raise the ceiling from 150 to 220 girls. The disappointed schools were then offered one and then a second date in July.

The format of the day was determined by our two main aims: to convey a girl-friendly message about the nature of mathematical activity, and to reinforce the narrowing effect on career prospects of disengaging from school mathematics.

The programme offered plenary sessions to open and close the conference; but most of the girls' time was spent in mathematics workshops or at a careers fair.

We sent participants details of the twelve workshops. They had the opportunity to attend two during the day and were asked to select the four that appealed most. On offer were sessions on Logo, 3-D models, probability, movement, calculators, games, puzzles and the maths of fashion and fabric. Amongst the leaders were three men.

This choice brought organisational problems. As the completed choice forms arrived we realised that too many girls were opting for the same few workshops. It seemed that an accurate description of a 'worthy' session was not enough to inspire. We could have filled the fashion workshop or the one on Logo four times whilst others offering investigations were not so popular. It seemed that the selling points were references to fashion, working with a friend and anything to do with micro-computers.

Eventually all the girls were allocated to two workshops each in such a way that everyone had a first or second choice.

All girls would also come to the careers fair to meet women doing non-traditional jobs. Their careers would not be identified as mathematical, but did require an 'O' level. They included an electronic engineer, a furniture designer, a British Telecom engineer, a quantity surveyor, a laboratory technician, and the head of TV cameras at the BBC (a male, looking for trainees).

During the fair each woman sat under a large label saying 'I am a ...' and the girls were free to walk around the Hall and talk to them informally.

A camera team from Chelsea College made a video of the activities which is to be used to support further school-based initiatives.

It was clear from the list of participants that the day would widen horizons irrespective of our input, because the schools ranged from selective suburban girls' schools to inner city comprehensives. Some teachers had brought a whole mixed-ability class, others a 'set' or a self-selected group.

There was some disappointment amongst those who had not been allocated to both their first and second choices of workshops; but, despite this, almost all the girls reported at the end of the day that they had enjoyed what they had done.

REACTIONS

Asked to complete the sentence beginning 'The best thing about today was ...', girls wrote: (The quotes are from the evaluation forms which the girls completed at the end of the day.)

> *learning about triangles. We learnt a lot and it was fun. We made shapes and talked about them.*
> *workshop 9 (Imagination, movement and maths) because everybody had the opportunity to get involved and it wasn't formal.*
> *my first workshop ... which was to do with shapes and we made a hexaflexagon and we tied ourselves up and we then had to try to get free.*
> *making models in workshop 5. The teachers were also very nice.*
> *the two workshops I went to. I was surprised how easy designing is with maths.*
> *the workshops. They have helped me learn that there is more to maths than sums.*
> *everything I done was good and interesting I much liked the logo I thought it was brilliant I would like to come again.*
> *computers, knitting and that's all.*

The response to the careers fair was mixed. The organisation relied upon the girls being sufficiently self-motivated and self-reliant to move around and ask questions. For a large proportion of them this was outside their experience and they absented themselves, later reporting the session as boring and badly organised. For others it was the best part of the day:

> *meeting women doing men's jobs and vice-versa was very good and enjoyable.*
> *the best thing about today was going round talking to women whose jobs we may not think of doing.*

To our surprise, however, the conversations that the girls held with the women rarely focused on mathematics. They wanted to talk about the practical problems of working. 'What do they do with their children all day?' 'How do they manage the housework?'

One of the women described swift exits whenever she steered the conversation towards the importance of mathematics to her job. Another, however, had the opposite experience.

A powerful piece of information was an OHP slide listing the careers for which 'O' level mathematics is an entry requirement.

It is clear that the careers fair did need more structure and that several organisational changes must be made for the re-runs but the overall evaluation was enthusiastic, typified by:

> *Today was beneficial and we're surprised how much maths has to do with everyday life.*
> *I enjoyed it the people were really nice and I would love to come again. It was a different experience...*

We had not thought that the venue was very important but for many girls it was clearly an unusual experience to be in a pleasant environment. They enjoyed walking in the grounds, winter gardens and the library between sessions. They were also trusted to find their way around a large college and to be in the right place at the right time.

It was an opportunity to meet girls of the same age but from very different schools. They were far less intimidated by this than might be expected and many reported 'making new friends' as one of the best things about the day.

They had been asked to make choices about the sort of mathematics they wanted to do. Opting-in (even if it wasn't a first choice!) was important to the atmosphere of the workshops.

The workshops were run by people without the teacher label. Several girls remarked that it felt so much better to know and use the first name of the people they worked with.

> *The staff made a nice atmosphere*

Their teachers weren't there! This was a contentious point with the teachers. Each school party was accompanied by at least one member of staff and a separate programme was arranged for them. This annoyed some of the teachers (and girls), who would have liked to see the activities but we believe that their presence would have inhibited the girls.

The feeling in the workshops was very relaxed. Each leader used her or his own mechanisms to achieve this but all shared the determination that everyone should be active, involved and unthreatened.

The girls were responsible for the discipline of the group. It would be gratifying to report that there were no behavioural problems. Recorded on video-tape is the evidence to the contrary. To quote:

> *there are some very selfish people about who pull out the computers when others want to use them and lose other people's information.*

The incident was resolved with virtually no action from the workshop leader.

Mathematics was redefined. The most striking examples of this come from the workshop on fashion and fabric. Here girls worked on complex problems of sleeve-fitting, fabric-matching and pattern-cutting and were confronted with the reality that they had been working on mathematics.

Mathematics was assumed to be worthy of protracted activity. One group of girls said they had not thought they could work for a whole day on mathematics, but now they wished there had been more time available. Many asked for four workshops not two or said that each session had been too short.

The 'problem' was brought out of the closet! Some girls said that maths was not a problem for them because they were at a girls' school or because they 'did SMILE' (Secondary Mathematics Individualised Learning Experiment, developed by ILEA).

Some girls were free from male-dominated classrooms for the first time. This was particularly striking in workshops using microcomputers where there were vivid descriptions of boys denying access to the girls.

> *The best thing about today was that there were no boys and no-one got into trouble.*

Girl-friendly mathematics was seen as possible. As one girl wrote three days later:

> *I don't know what you did with her but Miss X has become much nicer.*

REFLECTIONS

It is our firm conviction that sex-differentiation in mathematics education reflects the distribution of power in our society. The school and the classroom, and indeed very often the family, are part of and mirror that distribution. A female prime minister is no evidence of a re-distribution of political power. One only has to examine the numbers and roles of women in the Cabinet, in the Houses of Commons and Lords, in the Civil Service, in the top echelons of industry and, to come nearer to home, in headships and senior management in schools, to be convinced that power and decision-making still rest with men. In ILEA secondary schools, 51 per cent of teachers are full-time women of whom approximately 60 per cent are on scales 1 and 2. In primary schools, 80 per cent of teachers are women; 90 per cent of scale 1 teachers are women. Only 57 per cent of head teachers in these schools are women. And ILEA is a better Authority in this respect that most. (Statistics supplied by Research and Statistics Branch, ILEA.)
Examine the sex distribution of authority in your school/governing body/inspectorate/LEA administrators. (In this section we adapt the 'Cockcroft convention' and indicate with bold print suggestions for action.)

Under such circumstances, male styles of working, male preferences and male-identified high-priority subjects are bound to achieve higher status and a greater propensity to operate. However, the economic facts are that many families are currently dependent upon women's earnings. The 1974 Family Expenditure Survey showed that without the contribution of women's earnings, the number of families living below supplementary benefit level would have trebled (quoted in Marland, 1983) and times have surely worsened since then. In these circumstances, we, as teachers, have a great responsibility to our girl pupils to ensure that, at the least, they understand and are aware of the factors which can operate to their advantage or disadvantage. **You might initiate discussions with your careers colleagues** to ensure that they are aware of the critical filter-role of mathematics and that this awareness is passed to the pupils. Simultaneously, you might **look at currently used, successful techniques for raising awareness** of the issues and changing beliefs held by parents, colleagues and pupils. You could, for example, **focus on the stereotyping of mathematics as a male domain** by showing a videotape of the *Horizon* programme, *Mathematical Mystery Tour*, and concentrating on the roles women play in it and the hidden messages of the video about women's contribution to mathematics. You could **research information about women mathematicians** (see, for example, an excellent pack developed for schools in Victoria, Australia, on Women in Maths and Science) and point out how their contribution has disappeared from the history of mathematics. **Examine the textbooks and resources** you use for sex-stereotyping. This is particularly apparent in the illustrations of primary schemes and the impressions given by distributors of resources that active, scientific and enquiry-based equipment is most appropriate to boys. **Investigate the feelings of your pupils towards mathematics** by putting them into groups and asking half the groups to discuss and report on the subject of the following anecdote:

'John won a prize for mathematics last term. Describe John!'

Ask the other groups to do the same but with Anne as the subject.

Attitudes towards mathematics can be tested by attitude questionnaires which can then be discussed openly in the classroom. Evidence suggests that a fear of success operates amongst girl mathematicians depressing their mathematical performance and often leading them to deny interest or ability in an area which is conventionally identified as male (Horner, 1972). In addition, there is substantial evidence that boys and girls react differently to success and failure, boys attributing success in mathematics to ability, girls to effort, boys attributing failure to lack of effort or bad luck, and girls to lack of ability. It is noticeable that girls' success is attributable by them to external factors whereas their failure is internalised. For boys this pattern is reversed. Teachers who are aware of these dangers can **try to avoid reinforcing the attribution patterns** by making factual statements and avoiding judgmental ones.

When girls are given the space to listen and exchange with one another in a collaborative way they often change their attitudes to mathematics. When a boy is a member of the group, there is a tendency for him to hijack control of the group and dominate it. This is consistent with the evidence that teachers' responses to boys are significantly more numerous than those to girls and that this is partly due to the demanding behaviour of certain boys who dominate the interactional exchanges. This is observed whatever the age of the children, from the nursery upwards. **Look at the use of the computer by girls and boys**. Who is controlling the action in a mixed-sex group? If the calculators are handed out, who uses them? Changing the pattern of interactions from class or individually-based to group-based removes the possibility of class dominance from an individual pupil and places pupils together into a situation where they need to develop the skills of working together, listening and evaluating what each member of the group says or does. These skills, in our society, are regarded as female. 'Those who have their language, their experience and themselves rejected ... seek protection in silence. Those who have their experience, their language and themselves validated ... are encouraged to assert themselves even more' (Spender 1983, p. 114). In single-sex classrooms, these differences are still noticeable and just as reflective of different experience and different feelings about the subject-matter. **Who is silenced and who is validated in your classroom**? Ask a colleague to observe.

There are many ways in which sensitivity to sexism in the mathematics classroom can be developed. We have only referred to a few here but an in-service pack is in preparation to support teachers who want to focus on these issues. We would only want to conclude by saying that it is our experience that once you begin to notice sexism, you cannot stop and the will to do something about it is generated. In the process of removing or challenging power, some people are going to feel their loss, and resist. That is to be expected and strategies must be developed for dealing with such resistance. We append some references which we have found helpful in establishing some facts in place of myths. But most of all, we recommend that you talk to your girl pupils and hear how they think it is and work together with them to change it.

REFERENCES

Brush, L. (1980) *Encouraging Girls in Mathematics: the Problem and the Solution*. Abt Books.

Cockcroft, W.H. (Chair) (1982) *Mathematics Counts*, Report of the Committee of Inquiry into the Teaching of Mathematics in Schools. London: HMSO.

Deem, R. (1980) *Schooling for Women's Work*. London: Routledge and Kegan Paul.

Delamont, S. (1980) *Sex Roles and the School*. London: Methuen.

Eddowes, M. (1983) *Humble Pi*. Harlow: Longman for Schools Council.

Horner, M.S. (1972) 'Towards an understanding of achievement-related conflicts in women', *Journal of Social Issues*, **28**, 157–75.

Marland, M. (1983) *Sex Differentiation and Schooling*. London: Heinemann.

Spender, D. (1983) 'Telling how it is: language and gender in the classroom' in Marland, M., *Sex Differentiation and Schooling*. London: Heinemann.

Walkerdine, V. and Walden, R. (1982) *Girls and Mathematics: The Early Years*, Bedford Way Papers 8. University of London Institute of Education.

Weiner, G. (1985) *Just a Bunch of Girls*. Milton Keynes: Open University Press.

16

*The Logic of Problem Generation: from Morality and Solving to De-Posing and Rebellion**

STEPHEN I. BROWN

TWO CULTURES

A quarter of a century ago, C.P. Snow accurately pointed out how little the two cultures—roughly the sciences and the humanities—have learned to understand each other and to gain from the wisdom they each have to offer.

> Between the two a gulf of mutual incomprehension—sometimes ... hostility and dislike, but most of all lack of understanding [emerges]. They have a curious distorted image of each other ... Non-scientists tend to think of scientists as brash and boastful ... [They] have a rooted impression that the scientists are shallowly optimistic, unaware of man's condition. On the other hand, the scientists believe that the literary intellectuals are totally lacking in foresight, peculiarly unconcerned with their brother men, in a deep sense anti-intellectual, anxious to restrict both art and thought to the existential moment.
>
> (Snow 1959, p. 12)

Not only are their problem solving styles different, but more importantly there are divergent views on what it means for something to be solved. It is worth observing that as a profession, mathematics education is almost by definition bound to the schizophrenic state of searching for and creating the 'snow-capped' bridges; for mathematics is more closely aligned with the culture and world view of science and education with the humanities.

*An expanded version of this chapter first appeared in *For the Learning of Mathematics*, **4** (1) February 1984. Reproduced with permission.

As we search for a better understanding of what problem solving might be about, however, we have not only neglected to build bridges, but we have tended to ignore most non-mathematical educational terrain that might be worth connecting in the first place.

In particular, we have overlooked those educational efforts in other fields which have been concerned with problem solving but have indicated that concern through a different language. Dewey's analysis of 'reflective thought' and of the concept of 'intelligence' would seem to offer a rich complement to much of the problem solving rhetoric. The role of *doubt*, *surprise* and *habit* in problem solving explored by Dewey would seem to complement much of the influential work of Polya, and would offer options we have not yet incorporated in much of our thinking about problem solving in the curriculum (Dewey 1920, 1933).

We have much to learn about the role of dialogue in problem solving, something we in mathematics education have tended to view in pale 'discovery exercise' terms at best. Yet the use and analysis of dialogue in educational settings has been the hallmark not only of English education, but of several curriculum programmes in other fields as well. 'Public controversy' in the social studies in the late 60s and early 70s was a central theme around which students were taught not only to carry on intelligent dialogue, but more importantly to unearth and to discuss controversial and sometimes incompatible points of view (Oliver and Newman 1970). It would enrich considerably what it is we call problem solving in mathematics, if we were to entertain the possibility that for logical as well as pedagogical reasons, we might encourage not merely complementary, but incompatible perspectives on a problem or a series of problems. Furthermore such curriculum in the social studies as well as in the newly emerging field of philosophy for children might enable us to help students *appreciate* irreconcilable differences rather than to resolve or dissolve them as we are prone to do in mathematics (Lipman *et al.*, 1977).

'Critical' thinking is another 'near relative' of problem solving that began influencing the curriculum in schools as far back as the progressive education era, and there is a considerable history of efforts to integrate different disciplines through the use of critical thought (Taba 1950). It is a history that is worth understanding not only because of its connection with problem solving, but because the theme is presently undergoing rejuvenation in the non-scientific disciplines much as problem solving has re-emerged in mathematics and science.

MORAL EDUCATION AND KOHLBERG

One area within which the tunes of critical thinking have been re-sung recently is that of moral education. We might ask why critical thinking and moral education have been joined at all. To many people, they would seem to occupy different poles. The connection hinges on our concern for the teaching of values in a pluralistic, democratic society. How do we go about such education in a public school setting without indoctrinating with regard to a particular religious or ethnic point of view? Though we might argue over whether or not *it* is a set of values itself and if so, why it is that such a collection is more neutral than any religious or ethnic point of view, the liberal tradition of thinking critically about whatever values one adopts does provide an entrée for those concerned with morality in a pluralistic society.

Though there are a number of different kinds of programs within which moral education is taught (Lickona 1976), most of them rely heavily upon contrived or natural dilemmas as a starting point. Our focus here will be on Lawrence Kohlberg's program of moral development and education. A typical dilemma he has used for much of his research and for his deliberate program of education as well is the Heinz dilemma:

> In Europe, a woman was near death from a rare form of cancer. There was one drug that the doctors thought might save her, a form of radium that a druggist in the same town had recently discovered. The druggist was charging $2000, ten times what the drug cost him to make. The sick woman's husband, Heinz, went to everyone he knew to borrow the money, but he could only get together about half of what the drug cost. He told the druggist that his wife was dying and asked him to sell it cheaper or let him pay later. But the druggist said, 'no.' So Heinz got desperate and broke into the man's store to steal the drug for his wife.
> (Kohlberg 1976, p. 42)

Should Heinz have stolen the drug? Based upon an analysis of longitudinal case studies to answers of dilemmas of this sort, Kohlberg has created a scheme of moral growth that he claims is developmental. Furthermore, he has created not only a research tool but an educational programme around such dilemmas. It is through discussing and justifying responses to such dilemmas that students mature in their ability to find good reasons for their choices.

It is not the specific value that one chooses (e.g. steal the drug *versus* allow the wife to die), but the reasons offered for the decision that places people along a scale of moral development.

At the lowest level of moral maturity (pre-conventional) Kohlberg finds that people argue primarily from an awareness of punishment and reward. Thus someone at a lowest stage of development might claim that Heinz should not steal the drug because he would be punished by being sent to jail, or he might claim that he should steal it because his wife might pay him well for doing so. It is almost as if the punishment inheres in the action itself. At a later stage (conventional) people argue from the more abstract perspective of what is expected of you and also from the point of view of the need to maintain law and order. At the highest stage of principled morality, one argues on the basis not of rules that could conceivably change but with regard for abstract principles of justice and respect for the dignity of human beings. Such principles single out fairness and impartiality as part of the very definition of morality.

None of these structural arguments (e.g. punishment/reward, law and order, justice) in themselves dictate what is a correct resolution of any dilemma. Rather they form part of the web that is used to justify the decisions made, and it is in listening to these reasons that Kohlberg and his followers are capable of deciding upon one's level of moral development.

Gilligan's challenge

Despite the fact that Kohlberg's scheme for negotiating moral development neglects to focus upon action, it is a refreshing counterpoint to a programme of moral education which conceives of its role as one of inculcating specific values in the absence of reason. Nevertheless, there has been some penetrating criticism of his scheme recently—a criticism which condemns much of Kohlberg's work on grounds of sexism. That is, Kohlberg's research and ultimately his scheme for what represents a correct hierarchy of development is based upon his longitudinal research *only with males*. Once the scheme was created and the stages developmentally construed, Kohlberg interviewed females and concluded that their deviation from the established hierarchical scheme implied an arrested form of moral development.

Carol Gilligan (1982) points out that the existence of a totally different category scheme for men and women not only may be a consequence of different psychological dynamics, but rather than exhibiting a logically inferior mind set, it suggests moral categories that are desperately in need of incorporation with those already derived. Compare the following two responses to the Heinz dilemma, one by Jake, an eleven-year-old boy and the second by Amy, an eleven-year-old girl. Jake is clear that Heinz should steal the drug at the outset, and justifies his choice as follows:

> *For one thing a human life is worth more than money, and if the druggist makes only $1000, he is still going to live, but if Heinz doesn't steal the drug, his wife is going to die. (Why is life worth more than money?) Because the druggist can get a thousand dollars later from rich people with cancer, but Heinz can't get his wife again. (Why not?) Because people are all different and so you couldn't get Heinz's wife again.*
> *(Gilligan 1982, p. 26)*

Amy on the other hand equivocates in responding to whether or not Heinz should steal the drug:

> *Well, I don't think so. I think there might be other ways besides stealing it, like if he could borrow the money or make a loan or something, but he really shouldn't steal the drug—but his wife shouldn't die either. If he stole the drug, he might save his wife then, but if he did, he might have to go to jail, and then his wife might get sicker again, and he couldn't get more of the drug, and it might not be good. So, they should really just talk it out and find some other way to make the money.*
> *(Gilligan 1982, p. 28)*

Notice that Jake *accepts* the dilemma and begins to argue over the relationship of property to life. Amy, on the other hand, is less interested in property and focuses more on the interpersonal dynamics among the characters. More importantly, Amy refutes to accept the dilemma as it is stated, but is searching for some less polarised and less of a zero sum game.

Kohlberg's interpretation of such a response would imply that Amy does not have a mature understanding of the nature of the moral issue involved—that she neglects to appreciate that this hypothetical case is attempting to test the sense in which the subject appreciates that in a moral scheme life takes precedence over property. Gilligan on the other hand in analysing a large number of such responses has concluded not that the females are arrested in their ability to move through his developmental scheme, but that they tend to abide by a system which is orthogonal to that developed by Kohlberg—a system within which the concepts of *caring* and *responsibility* rather than *justice* and *rights* ripen over time.

Gilligan comments with regard to Amy's response:

> Her world is a world of relationships and psychological truths where an awareness of the connection between people gives rise to a recognition of responsibility for one another, a perception of the need for

> response. Seen in this light, her understanding of morality as arising from the recognition of relationship, her belief in communication as the mode of conflict resolution, and her conviction that the solution of the dilemma will follow from its compelling representation seem far from naive or cognitively immature.
>
> (Gilligan 1982, p. 30)

The difference between a 'Kohlbergian' and a 'Gilliganish' conception of morality is well captured by two different adult responses to the question, 'what does morality mean to you?' (Lyons 1983). A man interviewed comments:

> Morality is basically having a reason for doing what's right, what one ought to do; and, when you are put in a situation where you have to choose from amongst alternatives, being able to recognize when there is an issue of 'ought' at stake and when there is not; and then ... having some reason for choosing among alternatives.
>
> (Lyons 1983, p.125)

A woman interviewed on the same question comments:

> Morality is a type of consciousness, I guess a sensitivity to humanity, that you can affect someone else's life. You can affect you own life and you have the responsibility not to endanger other people's lives or to hurt other people. So morality is complex. Morality is realizing that there is a play between self and others and that you are going to have to take responsibility for both of them. It's sort of a consciousness of your influence over what's going on.
>
> (Lyons 1983, p. 125)

While Gilligan and her associates do not claim that development is sex bound in such a way that the two systems are tightly partitioned according to gender, they do claim to have located a scheme that tends to be associated more readily with a female than a male voice. Behind the female voice of responsibility and caring, some of the following characteristics appear to me to surface:

(a) a context-boundedness;
(b) a disinclination to set general principles to be used in future cases;
(c) a concern with connectedness among people.

Though not all of these characteristics are exhibited in Amy's response, they do appear in interviews with mature women. *Context-boundedness* represents a plea for more information that takes the form not only of requesting more details (e.g. what is the relationship between husband and wife?) but of searching for a way of locating the episode within a broader context. Thus unlike men, mature women might tend to respond not by trying to resolve the dilemma, but by exhibiting a sense of *indignation* that such a situation as the Heinz dilemma might arise in the first place. Such a response might take the following form: 'The question you should be asking me is "What are the horrendous circumstances that caused our society to evolve in such a way that dilemmas of this sort could even arise—that people have to miscommunicate so poorly"?'

The second characteristic I have isolated above, is an effort to attempt to understand each situation in a fresh light, rather than in a legalistic way—i.e. in terms of already established precedent. Connected with context-boundedness it is the desire to see the fullness of 'this' situation in order to see how it might be *different from* (and thus require new insight) rather than compatible with one that has already been settled.

With regard to the third characteristic, conflict is less a logical puzzle to be resolved but rather an indication of an unfortunate fracture in human relationships—something to be 'mended' rather than an invitation for some judgement.

In the next section we turn towards a consideration, in a rather global way, of how it is that a Gilliganish perspective of morality might impinge on the study of mathematics. While we have not yet drawn any explicit links, it is not difficult to intuit not only that it threatens the status quo but that it sets a possible foundation for the relationship of problem generation to problem solving. Though we shall focus upon the findings from the field of moral education, we do not wish to lose sight of some of the other humanistic areas of curriculum from which mathematics education might derive enlightenment.

KOHLBERG *VERSUS* GILLIGAN: THE TRANSITION FROM SOLVING TO POSING

It surely appears that problem solving in mathematics education has been dominated by a Kohlbergian rather than a Gilliganish perspective. Gilligan herself has an intuition for such a proposition, when she comments with regard to Jake's response to the Heinz dilemma:

> Fascinated by the power of logic, this eleven-year-old boy locates truth in math, which he says is 'the only thing that is totally logical.' Considering the moral dilemma to be 'sort of like a math problem with humans,' he sets it up as an equation and proceeds to work out the solution. Since his solution is rationally derived, he assumes that anyone following reason would arrive at the same conclusion and thus

> that a judge would consider stealing to be the right thing for Heinz to do.
>
> (Gilligan 1982, p. 26–7)

The set of problems to be solved as well as the axioms and definitions to be woven into proofs are part of 'the given'—the taken-for-granted reality upon which students are to operate. It is not only that the curriculum is 'de-peopled' in that contexts and concepts are for the most part presented ahistorically and unproblematically, but as it is presently constituted the curriculum offers little encouragement for students to move beyond merely accepting the non-purposeful tasks.

Furthermore, rather than being encouraged to try to capture what may be *unique* and unrelated to previous established precedent in a given mathematical activity (the legalistic mode of thought we referred to as the second characteristic behind Gilligan's analysis of morality as responsibility and caring), much of the curriculum is presented as an 'unfolding' so that one is 'supposed' to see similarity rather than difference with past experience. It is commonplace surely in word problems to tell people to *ignore* rather than to embellish matters of detail on the ground that one is after the underlying structure and not the 'noise' that inheres in the problem.

In so focusing on essential isomorphic features of structures, the curriculum tends not only to threaten a Gilliganish perspective, but as importantly, it supports only one half of what I perceive much of mathematics to be about. That is, mathematics not only is a search for what is essentially common among ostensibly different structures, but is as much an effort to reveal essential differences among structures that appear to be similar (See Brown, 1982a).

With regard to context-boundedness, there is essentially no curriculum that would encourage students explicitly to ask questions like:

1. What purpose is served by my solving this problem or this set of problems?
2. Why am I being asked to engage in this activity at this time?
3. What am I finding out about myself and others as a result of participating in this task?
4. How is the relationship of mathematics to society and culture illuminated by my studying how I or other people in the history of the discipline have viewed this phenomenon?

Elsewhere (Brown 1973, 1983) I have discussed how I first began to incorporate such reflection as part of my own mathematics teaching, and presently I shall have other illustrations. There are a number of serious questions that must be thought through, however, before one feels comfortable in encouraging the generation and reflection of the kinds of questions indicated above. We need to be asking ourselves whether or not that kind of reflection represents respectable mathematical thinking. In addition we ought to be concerned about the ability of students to handle that thinking in their early stages of mathematical development.

It is interesting to observe that though we are cajoled by many to integrate mathematics with other fields, the 'real-world' applications seem to be narrowly defined in terms of the scientific rather than the humanistic disciplines. In particular, questions of value or ethics are essentially non-existent. That is particularly surprising in light of the fact that a major rationale for relating mathematics to other fields seems to be that such activity may enable students to better solve 'real-world' problems that they encounter on their own. I know of essentially no 'real-world' problems that one decides to engage in for which there is not embedded some value implications.

McGinty and Meyerson (1980) suggest some steps one might want to take to develop curriculum for which value judgements are an explicit component. Consider a problem like the following:

> Suppose a bag of grass seed covers 400 square feet. How many bags would be needed to uniformly cover 1850 square feet?
>
> (McGinty and Meyersen 1980, p. 501)

So far so dull. It is not only that for many students the above would not constitute a problem, but more importantly it lacks any reasonable conception of context-boundedness. The authors, however, go on to suggest inquiry that is more real-worldish that most of the word problems students encounter. They ask:

> Should the person buy 5 bags and save the leftover—figuring prices will rise next year? Buy 5 bags and spread it thicker? Buy 4 bags and spread it thinner?
>
> (McGinty and Meyerson 1980, p. 502)

Once we become aware of ethical/value questions as a central component of decision making, it is clear that there is much more we might do in the way of generating problems for students as well as encouraging them to do so on their own. One of the *au courant* curriculum areas is probability and statistics. As a profession, we correctly appreciate that we need to do more to prepare students to operate in an uncertain world, wherein one's fate is not sown with the kind of exactitude that much of the earlier curriculum has implied. In creating such a curriculum, however, we continue to give the illusion that mathematical competence is all that is required to decide wisely. Compare *any* probability problem (selected at random of course) from any curriculum in mathematics with the following probability problem:

> A close relative of yours has been hit by an automobile. He has been unconscious for one month. The doctors have told you that unless he is operated upon, he will live but remain a vegetable for the rest of his life. They can perform an operation which, if successful, would restore his consciousness. They have determined, however, that the probability of being successful is .05, and if they fail in their effort to restore consciousness, he will certainly die.

What counsel would you give the doctors? One could clearly embed the above problem in a more challenging mathematical setting, for example, setting up the conditions that would have enabled one to arrive at the .05 probability (or perhaps modifying it so that outer limits are set on the probability of survival) but nevertheless, it is such ethical questions in many different forms that plague most thinking people as they go through life making decisions.

Is such a problem-generation on the part of the teacher or student an ingredient of mathematical thought? I do not think the answer is clear. There is nothing god-given and written in stone that establishes what is and is not part of the domain of mathematics, and clearly what has constituted legitimate thinking in the discipline has changed considerably over time. Even if questions of the kind we have been raising in this section, however, would move us in directions that are at odds with the dominant and respectable mode of mathematical thought, it is worth appreciating that as educators we have a responsibility to future citizens that transcends our passing along *only* mathematical thought. The latter appears to me to be a very narrow view of what it means to educate. In realising that only a very small percentage of our students will be mathematicians, we have not adequately explored our obligation to those who will not expand the field *per se*. We have mistakenly identified our task for the majority as one of 'softening' an otherwise rigorous curriculum. What may be called for is an ever more intellectually demanding curriculum, but one in which mathematics is embedded in a web of concerns that are more 'real-world' oriented than any of us have begun to imagine.

DOWN FROM A CRESCENDO

The confrontation between Kohlberg and Gilligan has served two purposes that appear on the surface to be very different. First of all, we have used the challenge of Gilligan's research to point out that there is a world view that has achieved empirical expression with regard to issues of morality but which is worth taking seriously in other domains as well. Moving beneath the concepts of caring and responsibility established by Gilligan, we find dimensions that are not strictly moral in character but which deal with *purpose, situation specificity* (a non-legalistic mode) and *people connectedness*. We have suggested that very little of the existing mathematics curriculum caters to those characteristics, and in fact the dominant mode caters to their opposite.

Secondly, we have not only used Gilligan in contrast to Kohlberg to establish broad categories within which the present curriculum is deficient, but we have pointed out that what the two perspectives have in common—namely a concern with morality—represents a field of inquiry that may be as important to integrate with mathematical thinking as are the more standard disciplines that form the backbone of more conventional applications.

Both of these perspectives have potentially revolutionary implications. They not only suggest the need for both teacher and student to incorporate a more serious problem-generating perspective (including the broad types of questions raised at the beginning of the previous section) as an essential ingredient of problem solving, but they have the potential to infect every aspect of mathematics education from drill and practice, to an understanding of underlying mathematical structures.

Our goal for the remainder of this chapter will be the more modest one of making a case for the inclusion of problem-generating strategies within the curriculum. I will for the most part be drawing upon and integrating ideas that I have previously developed. While I will make minimal explicit reference to the Gilligan perspective, I believe it is possible to view much of what follows as being derived from what I have referred to as the underlying components of caring and being responsible. The joining of links explicitly in other mathematics education areas is a task to be left for another time (and perhaps another person).

OUR KNEE-JERK SOLVING MENTALITY

Any field of inquiry establishes a common language among its investigators. The same kind of phenomenon is exhibited among friends, lovers and members of a family. It is frequently possible to determine the extent to which you are in fact an 'outsider' by the degree to which you are incapable of understanding the short-circuiting of language among participants. There is certainly good reason for members of an 'in-group' to engage in such short-circuiting behaviour. In addition to merely increasing efficiency of communication, there are important psychological and sociological bonds established through such behaviour.

Nevertheless, we sometimes pay a price for the common language we establish. That is, in focusing on common understanding, we not only leave out other perspectives, but we may be *unaware* of what we are leaving out. The specialisation that results from such behaviour not only may leave us unaware of what we have left out, but worse than that, we may even lose our ability to incorporate those awarenesses within our world view even when they are pointed out to us.

As educators, it is worth taking stock every so often to examine explicitly what we are leaving out in the common language we are establishing with our students. Such an instance occurred a number of years ago at which point Marion Walter and I were team-teaching a course on problem solving. We were doing work in number theory, and were hoping to derive a formula to generate primitive Pythagorean triplets. We began the lesson by asking:

$x^2 + y^2 = z^2$. What are some answers?

Responses began to flow, and students responded with:

3, 4, 5
5, 12, 13
8, 15, 17

After a while, a smile broke out on the face of a student who responded:

$1, 1, \sqrt{2}$

A few more 'courageous' and humorous responses then were suggested like:

$-1, -1, \sqrt{2}$

Marion and I then jokingly reprimanded the 'deviants', and proceeded to explore what we were about in the first place—a search for a generating formula. After class, however, we began to talk to each other about the incident. It hit us very hard that the 'deviants' were beginning to appreciate something that has occupied a considerable part of our collective energy for the past fifteen years. What struck us was that:

> $x^2 + y^2 = z^2$. What are some answers?

has a kind of foolishness about it that derives from the closed position of a common language with its unspoken but built-in assumptions. Notice that $x^2 + y^2 = z^2$ is not even a question. How can one come up with answers?

Yet the students dutifully did come up with answers, because they carried along a host of assumptions that we in fact have trained (implicitly) them to accept. They assumed (at least at the beginning) that the symbolism had connoted that the domain was natural numbers. Furthermore they assumed that the symbolism was calling for something algebraic; and within that context they assumed that we were searching for instances that would make an open sentence true.

As soon as we began to appreciate that 'the deviants' had begun to appreciate something we had not seen, we realised that there was a whole new ball game at stake. We had not realised at the time that in expanding this concept for this class, we were opening Pandora's box.

After realising that we had implicitly assumed that the domain was natural numbers, we encouraged students to ask such new questions as:

> For what *rational* numbers x, y, z is it true that $x^2 + y^2 = z^2$?

Realising that we had implicitly assumed that we were searching for true instances of the open sentence, we encouraged students to ask such new questions as:

> For what natural numbers is it true that $x^2 + y^2 = z^2$ is 'almost' true? (e.g. 4, 7, 8 misses the equality by 1).

Realising that we had implicitly assumed that the question was algebraic, the students began to ask a host of geometric questions that derived from connotations of the algebraic form.

What followed immediately was one of the most intellectually stimulating units that either of us had previously experienced with our students, and what dawned eventually on all of us was something that has had a lasting effect.

First of all, we began to appreciate that such deviations from standard curriculum are not mere frills. That is, in exploring such questions as the 'almost' primitive Pythagorean triplet question, all of us gained a much clearer understanding of what the actual primitive

Pythagorean triplet question was in fact about—not only from the point of view of statement but of proof as well.

Secondly, and more importantly, we began to realise that an implicit part of the common language we share with students is one which focuses upon and points so strongly towards the search for solutions and answers, that we continue to search for answers even when no question is asked at all! We were thus launched on our journey to try to understand the role of problem-generation in the doing of mathematics.

POSING AND DE-POSING: A FIRST STEP

I am beginning to appreciate an important aspect of what is behind an understanding of the role of posing problems that I have not seen before, despite the fact that I have referred in much of my writing to examples within which this issue is embedded. For a number of years, educators have appreciated that there might be considerable value in giving students not *problems to solve* but *situations to investigate*. Higginson (1973), for example, locates a number of characteristics of what he refers to as 'potentially rich situations'. Situations are much 'looser' than problems, and situations themselves do not ask built-in questions. It is the job of the student to create a question or pose a problem. Geoboards, Cuisenaire Rods, polyominoes are all examples of situations, but situations need not be concrete materials; they can be abstractions as well.

What I have recently begun to appreciate is that the pedagogical issue is much deeper and more interesting than that of directing teachers to create rich situations for their students to investigate. The issue is even more complicated than providing both mechanisms and an atmosphere within which problem solving might be isolated from situation-related activities. Rather the pedagogical task is one of enabling all of us to appreciate the *differences* between a problem and a situation, and of finding ways to move from one to the other.

The task of so moving is neither mechanical nor easy. That it sometimes takes a very long time to appreciate that a situation implies a problem is something that most parents experience through much of their child rearing. That problems can be neutralised (or de-posed as the title of this section playfully suggests) is something that may be equally difficult to appreciate. Those of us who realise that we have been asking the wrong questions realise implicitly the need to move from a problem to a situation before re-posing the problem.

Consider the example of a 'female response' to the Heinz dilemma which asserts with indignation that the problem is not one of stealing or not stealing the drug, but rather one of figuring out how we even evolved as a society such that such choices would have to be made (and one of figuring out how to reconstruct society). Here is a clear case of first neutralising a problem before re-posing it. It was necessary to delete the question (Should Heinz steal the drug?) before moving towards a re-posing of the problem.

It is not only that there is value in having students actually move in both directions—from situations to posing and from posing to de-posing—but it is also worth designing a curriculum which exhibits the difficulties people had in making such moves on their own in the history of discipline. We have the potential to learn a great deal about the relationship of a discipline to the culture from which it emerges as we study those problems that could not be perceived as situations.

An obvious example in the history of mathematics is that of efforts over several centuries to try to prove the parallel postulate. Consider the following formulation of the question:

> How can you prove the parallel postulate from the other postulates of Euclidean geometry?

We know now that a great deal of the history of mathematics was written as nineteenth century mathematicians began to appreciate that the difficulty in solving the problem was that a wrong question was being posed. In some implicit sense, Lobacheysky and his colleagues at the time had in fact to 'neutralise' the problem enough first to get clearly at the *situation* from which it derived (the postulates of Euclidean geometry) and then to reformulate the question so as to delete the deceptively innocent word 'How' in the posing of the problem.

The need to re-pose a problem by first neutralising it is not only revealed through frustrated efforts at solving problems, but is an aesthetic issue as well, and an issue that is worth incorporating explicitly in curricula within which the Gilliganish concept of context-boundedness is taken seriously. Consider the case of efforts to prove the four colour conjecture—roughly that for any conventional map, four is a sufficient number of colours to establish and appropriately to demarcate boundaries. Until recently the problem was 'merely' to prove or disprove that conjecture. Only after a computer proof was produced which featured a very large number of special cases did mathematicians begin to realise that they had not adequately posed the problem. Feeling that a computer proof was blind to underlying structure and in fact illuminated very little of 'the mathematical essence' of the problem, many mathematicians realised the need to state the problem in such a way that 'ugly' proofs would not count as solutions.

Such re-posing of the four colour problem reveals something not only about the present attitude of many mathematicians with regard to the computer, but just as importantly, it unearths some fundamental epistemological issues—issues that more clearly locate knowledge within an aesthetic realm.

From a pedagogical point of view, it is particularly enlightening to engage students in a discussion of the relationship of a situation to a problem. I have a modest example. Several years ago my son, Jordan, came to me to tell me that he did not understand the 'ambiguous case' in trigonometry, i.e., those circumstances under which a triangle is determined by an angle, another angle and a side not included between the angles.

I began my discussion with him by asking him to recall how in geometry, he had investigated those conditions under which a triangle was determined. Jordan looked very puzzled and told me that I was mistaken; they had never investigated the determination of a triangle. Instead they had proven things about two triangles being congruent if A.S.A. = A.S.A and so forth.

What was taking place here is very interesting from the point of view of relating a problem to a situation. Jordan had in fact viewed an entire unit of work more as a *situation*, while I had viewed it as a *problem*. That is, though he had an arsenal of congruence theorems at his disposal to respond to any request to prove two triangles congruent, he did not see this ammunition as providing answers to what I saw to be the fundamental problem of discovering those conditions under which a triangle is determined. As I reviewed his text, I understood why he saw a situation in what I saw to be a problem. The book had in fact never distinguished between an underlying problem (determining a triangle) and a collection of exercises to give one experience in handling a problem that had been solved by the famous congruence theorem. In fact, the practice exercises had become the fundamental concept—a phenomenon I am beginning to believe is more widespread than I had thought, and a consequence most likely of the essentially plagiaristic spirit that governs textbook writing.

The interesting irony in this case is that the difference between my perception and Jordan's regarding what those congruence theorems were all about, was not revealed in Jordan's performance in geometry at all. One can frequently accurately answer questions and even solve difficult problems without seeing the context within which those problems are embedded.

Thus, it would seem to be a very wise pedagogical ploy to move not only from situation to problem and back for topics that are relatively small (e.g. de-pose a theorem such as 'The base angles of an isosceles triangle are congruent'), but to do so for entire units as well. Teachers as well as students would find it enlightening to discover the areas of agreement and divergence of opinion regarding the problem/situation status of a unit or perhaps even of a course.

In closing this section, I would like to comment on an interesting potential difficulty relating situation to problem. Many people believe that proper selection of problems is critical in designing curricula for one does not want to give problems to students that they are not prepared to handle. This point of view appears on the surface to be a threat to the activity of posing and de-posing problems. That is, what happens if in the creation of a problem from a situation, a student defines a problem that we know is beyond his/her ability to handle?

There are some interesting assumptions embedded in the above question. First of all, it is not necessarily the case that students need to try to solve problems they pose. The activity of posing itself in the absence of efforts to solve may be illuminating both to students and teachers. In a sense we find out as much of value about ourselves by attending to the kinds of questions we ask as we do by the solutions we attempt.

Secondly, if we think of an entire class as a resource, for the kinds of activities suggested in this section, it is not necessarily the case that the same person who poses a problem need be obligated to try to solve it. In fact we may discover the potential for unexpected collaboration among those who pose and those who attempt to solve. We do not know very much at all about the relationship of the talent of posing and solving, but perhaps it is worth taking a clue from the work of Getzels and Jackson (1961) in which they find reason to conclude that beyond a certain point, intelligence and creativity may not be as closely related as one might suppose.

But there is another consideration that cuts deeper than those we have mentioned so far. That is, what do we imply students are incapable of doing when we say they are prematurely challenged? It appears that they may be incapable of *solving* problems that either we (as teachers) or they pose. But such an expectation may be a short-sighted one from an educational point of view. Along with our newly discovered appreciation for the role of approximation and estimation, ought to come an appreciation for partial solutions as a respectable activity. We need not necessarily expect a complete solution for every problem investigated. In addition, I am not clear on what it is that is lost if students attempt to solve a problem and cannot even come up with partial solutions. Suppose they cannot even identify or isolate lemmas that might help them along the way. I can imagine a great deal of valuable personal and intellectual insight that might emerge through a discussion of what may account for inability of students at a particular point in time to make headway in solving particular problems. A teacher who keeps an ear to the ground might possibly even learn something of the students' conception of the subject matter, proof, mathematics and the relationship of mathematics to culture by listening carefully to what counts as a reason for failure to make headway.

THE ACT OF POSING: LOGIC AND PEDAGOGY

In relating problem posing to the creation of situations, we have, beneath the surface, bumped up against the relationship of problem posing to problem solving. After all, it was due to an inability to solve the parallel postulate problem that a situation was revealed which was in need of reformulation. Problem generation and problem solving are intimately connected, however, even when things do not go awry. Below we discuss their intimate logical connection. In the two subsections that follow the one below we shall look more closely at pedagogical strategies for engaging in problem posing—one mild and the other radical. Much of what I will be analysing in this section has appeared in disparate sources, and I view the task here as one primarily of consolidating that material. For that reason this section will be briefer than the others and the reader's attention will be drawn to relevant references for expansion of the points alluded to.

Logical connections with solving

Consider the following two problems:

> (1) A fly and train are 15 km apart. The train travels towards the fly at a rate of 3 km/hr. The fly travels towards the train at a rate of 7 km/hr. After hitting the train, it heads back to its starting point. After hitting

the starting point, it once more heads back toward the train until they meet. The process continues. What is the total distance this fly travels?

(2) Given two equilateal triangles, find the side of a third one whose area is equal to that of the sum of the other two.

The first problem reveals in a dramatic way something that is true but less obvious in the solution of any problem. If you have not seen this problem before, let it sit for a whole, or perhaps share it with an eleven-year-old. If the wind is blowing properly, you will come upon an insight that will most likely jar and inspire you. Without giving the ball-game away completely, let me suggest that an insightful and non-technical solution depends upon your asking a question that has not been asked in the problem at all. Though there are many different ways of asking the question as well as many questions to ask, something like the following will most likely be revealing:

What do I notice if I focus not on the fly as requested, but on the train instead?

What is needed in the solution of this problem is some effort at posing a new problem within the context of accepting and trying to solve a given problem. Whether or not such a problem posing is *always* needed in the solution of a problem is an interesting and debatable question. I believe that such problem generation is always needed, but I also believe that the analysis of the assertion very much hinges on how it is that one defines a problem in the first place. (See Brown, 1981b and Brown and Walter 1983 for additional discussion of this point.)

The second problem reveals another interesting intimate connection between problem solving and problem generating. The solution depends (an illustration of what we have said above) upon how it is that the problem itself is redefined. If, however, you assume that sides and their lengths can be distinguished from each other (something that is *not* necessary in the solution of the problem), then if the lengths of the sides of the first two triangles are a and b respectively, we can prove without too much fanfare that the length of the third side c is equal to $\sqrt{(a^2 + b^2)}$.

Now in one sense we have solved the original problem. In another sense, however, we have only begun to solve it. Most people who come upon the solution, $c = \sqrt{(a^2 + b^2)}$ are taken aback. The point is that it smells as if this is an interesting and unexpected connection (as a matter of fact one which now enables one to solve the problem without associating the sides with their lengths). The fact that the

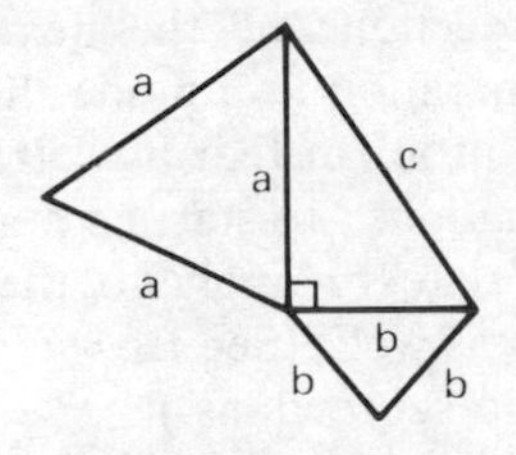

Figure 16.1

relationship is a Pythagorean one, indicates that we can find the third side as suggested in Figure 16.1.

Most mathematicians who have not seen this problem before find themselves headed in an almost compulsive search for what is happening. They are driven by some variation of the question:

> I know areas are additive for the squares on the sides of the right triangle, but *why* are they additive for equilateral triangles as well?

What this example illustrates very nicely is that a proof or a solution in itself does not always reveal *why* things operate as they do. Something more is needed, and in this case that something more begins with a question.

Though it is surely the case that the alleged solution of any problem always has further implications that one may assert as a problem or a question, one is not necessarily driven to do so in all problems with the same kind of fervour as in this case. (See Walter and Brown 1977, and Brown and Walter, 1983 for an elaboration of this discussion).

There are pedagogical implications that flow from these relationships between posing and solving problems. Students are not always aware of the questions they may have implicitly asked themselves in coming up with the solution to a problem, and there might be value in encouraging them explicitly to see what they have done. At the other end of the spectrum, students may not at all be aware of additional questions they 'need to' or might ask after they have supposedly solved a problem.

On strategies for posing: an accepting mode

It is one thing to suggest that problem posing is worthwhile, or even necessary; it is another to be able to do it. We shall in this subsection suggest several strategies for posing problems, some of which are well discussed in the literature, and some of which represent new directions. In this subsection and the next, we shall look at the activity of problem generation in a mode that is somewhat isolated from that of solving a problem that has already been stated. In so doing, we return to situations as a starting point. Much of what we do here might be appropriate to apply to the activity of solving an already stated problem as well. (See Brown and Walter, 1983 for an elaboration of these two subsections.)

What are the 'things' that situations are made of? Among possible candidates are the following:

1. *Concrete objects* like Cuisenaire rods and the Tower of Hanoi.
2. *Abstract 'things'* like
   (a) isosceles triangles; or
   (b) nine Supreme Court justices each shaking hands with each other.
3. *Data* like
   (a) primitive Pythagorean triplets generated by the relationships $x^2 + y^2 = z^2$; or
   (b) 5, 12, 19, 26, 33 ...
4. *Theorems or postulates* like the fundamental theorem of arithmetic (Every number can be expressed uniquely as a product of primes.)

There are surely more kinds of 'things' that one might use as a starting situation, but the above should serve the purpose of enabling us to see the directions we might look towards in generating questions.

(i) Estimation/approximation

Here is a category with which we are all familiar, though we tend not to make as much use of it in practice as we might. Given phenomenon 2(b) for example, most people with a little knowledge will ask: How many handshakes are there? Of course it is just as illuminating (for some purposes) a question to ask: *About* how many handshakes are there?

(ii) Internal and external views of a thing

Given situation 2(a), most people will ask the rather familiar question: What can you say about the base angles? Some people might extend the base and ask about the external angles. Compare those kinds of questions with one like:

> How many isosceles triangles can you join to form the hub for the bicycle wheel?

How does the above question differ from the other isosceles triangle questions? It is worth pointing out that while the first set focuses on the internal working of the phenomenon, the one dealing with the hub takes the isosceles triangle in its entirety and relates it to something else. Much of our standard curriculum is focused on an internal view of objects and relatively little takes as its starting point the object as a whole.

(iii) The particular and the specific

Here is a theme that is particularly salient in terms of a Gilligan perspective. Take a look at 3 above and pose some problems.

Our enchantment with abstraction and generalisability frequently blinds us to the uniqueness of what is before us. Most people shown 3(a) and (b) will pose a problem that attempts to reveal some covering law that will generate all the terms. A careful look at data, however, frequently suggests that there is more to see that might be equally as appealing. Consider the following for example with regard to 3(a):

> Each triplet has at least one member divisible by 3, by 4, and by 5. Will that hold in general?

The above is clearly *not* a question that would arise if our focus were upon the more abstract Pythagorean relationship.

Take another look at 3(b). What questions arise from a careful look at the data beyond a search for some general algebraic generating formula?

(iv) On pseudo-history

Many teachers wish they knew more about the history of mathematics so that they might be better able to motivate the subject. What is not well appreciated, however, is that a great deal of intellectually stimulating thought can flow from an effort on the part of students as well as teachers to engage in what I call pseudo-history (Brown 1978 as well as Brown and Walter 1983). As an example, consider the following kind of question conceivably generated by 4.

> What *might* have been responsible for getting people to look at products of primes.

We can, for example, imagine a mathematics community that focused originally on expressing any given number as the *sum* of other numbers. What might have moved them to look at *products* instead?

These are surely not the only categories for generating problems while at the same time maintaining an accepting view towards the beginning situation. They do, however, represent a start, and with the exception of (i) tend not to be given much curriculum consideration. It would be a valuable contribution to expand both the list of 'things' to use as starting points as well as the categories one might look towards in the generation of questions for each of these 'things'.

Posing as an adolescent

In this subsection we further expand the concept of problem generation by selecting a situation in a mode that is more reminiscent of adolescent rebellion than is the previous subsection.

The concept of challenge, threat or adolescent rebellion is well captured by Douglas Hofstadter (1982) when he comments:

> *George Bernard Shaw once wrote (in Back to Methuselah)*: 'You see things; and you say 'Why?' But I dream things that never were; and I say 'Why not?' When I first heard this euphuism, it made a lasting impression on me. To 'dream things that never were'—this is not just a poetic phrase but a truth about human nature. Even the dullest of us is endowed with this strange ability to construct counterfactual worlds and to dream. Why do we have it? What sense does it make? How can one dream or even 'see' what is visibly not there? ... Making variations on a theme is really the crux of creativity. On the face of it the thesis is crazy. How can it possibly be true? Aren't variations simply derivative notions, never truly original creation?
>
> (p. 20)
>
> Careful analysis leads one to see that what we choose to call a new theme is itself always some kind of variation, on a deep level of earlier themes.
>
> (p.29)

One can start with a definition, a theorem, a concrete material, data, or any other phenomenon and instead of *accepting* it as the given to be explored, one can challenge it and in the act create a new (in some sense) 'it'.

> Consider for example the definition of a prime number: A natural number is prime if it has exactly two different divisors.

Now the 'natural' inclination of the standard curriculum is to use that bit of information to prove or show all kinds of things. An adolescent rebellion on the other hand might generate a host of questions like:

> What's so special about numbers that have exactly *two* different divisors? What kinds of numbers have exactly three divisors?

> Why do we focus on *divisor*? Can we find numbers that have exactly two different elements to form a sum?
>
> Why are we focusing on *different* divisors? Can numbers have the *same* divisor twice?
>
> Why do we focus on *natural* numbers? Suppose we look instead at fractions or the set of odd integers.

I shall not continue with the list of such questions that can be generated to challenge rather than accept the concept of prime number. (See Brown, 1978 and 1981a, for a thorough development especially of the last question.) Let me merely indicate that such activity has a built-in kind of irony, for it is in the act of 'rebellion' that one comes to understand better the 'thing' against which one rebels. In that sense challenging 'the given' as a strategy for problem generating has the potential to be viewed as a less radical departure from standard curriculum than one might otherwise believe.

Marion Walter and I have taken the insight of challenging the given and created a scheme which we call 'what-if-not'. A number of people have derived some fascinating and imaginative concepts by employing the scheme. Though it is possible to approach that scheme in an overly mechanistic manner, it is also something that can be done with taste.

Suppose one wishes to do a 'what-if-not' on the Fibonacci sequence:

> 1, 1, 2, 3, 5, 8, 13, 21, 34, 55, ...

For the *first* stage of the scheme, one lists the attributes of 'the thing', without worrying about such matters as completeness, repetition, elegance of statement, independence of statements and so forth. Thus we might list among the attributes:

1. The sequence begins with the same first number.
2. The first two numbers are 1.
3. If we do *something* to any two successive terms, we get the next number in the sequence.
4. The something we do is add.

At the second stage, we do a 'what-if-not' (hence the name of the scheme) on one of the attributes. For example, suppose we do a 'what-if-not' on the second attribute above; if it is not the case that the first two numbers are 1, we ask what they might be? Obviously we could select many alternatives to 1 and 1 as the starting numbers. Suppose we chose 3 and 7.

At the next stage, we ask some new set of questions about the modified phenomenon. Suppose we begin by asking what the new sequence would look like. To continue the process, we finally engage in the kind of activity which most people incorrectly assume is the essence of mathematics—namely we analyse or try to answer the question. Thus, if we maintain the essential definition of the original sequence (something we need not necessarily feel obligated to do), we would get:

3, 7, 10, 17, 27, 44, 71, ...

Moving back to the stage of asking some new set of questions, we might ask:

1. Is there an explicit formula to generate the *n*th term of the sequence?
2. How do properties of this sequence compare with those of the original one?

An analysis of these questions reveals some *very* fascinating jewels. People who are familiar with properties of the original Fibonacci sequence, in analysing the second question above, most likely would look (among other things) at ratios of succeeding terms. Choosing smaller to larger adjacent terms, we would get:

0.42, 0.70, 0.588, 0.708, 0.614, 0.62

Something smells (as in the equilateral triangle example in the previous subsection) peculiar. We are arriving at ratios that appear to be very close to the 'golden ratio' (approximately 0.618)—something we expect from the original Fibonacci sequence. Why is that happening?

In analysing the question above, one is thrown back towards an analysis of the original phenomenon—as we indicated above.

We have barely begun to see the wealth of surprising results in making use of the 'what-if-not' strategy of the Fibonacci sequence. (See Brown 1976, and Brown and Walter 1983 for a more detailed discussion.) In this brief sketch, however, we implied the value both of carefully employing the various stages of the 'what-if-not' strategy and of interrelating them as well.

In closing, it is worth pointing out that despite efforts to mechanise the stages, the process of 'what-if-not'ing tends to elude a computerised mentality, for it is frequently the case that in the absence of an essentially human activity one may never even 'see' some of the attributes to vary in the first place. Elsewhere (Brown,

1971, 1974, 1975, 1981a, 1982b) I have shown how it is that use of poetic devices such as metaphor and imagery, and such human qualities as finding surprise and flipping figure and ground, frequently account for our ability to see what it is that is supposedly staring us in the face all along.

REFERENCES

Brown, Stephen I. 'Rationality, irrationality and surprise,' *Mathematics Teaching*, No. 55, Summer 1971, pp. 13–19.

Brown, Stephen I. 'Mathematics and humanistic themes: sum considerations,' *Educational Theory*, Vol. 23, No. 3, 1973, pp. 191–214.

Brown, Stephen I. 'Musing on multiplication', *Mathematics Teaching*, No. 61, December 1974, pp. 26–30.

Brown, Stephen I. 'A new multiplication algorithm; on the complexity of simplicity'. *The Arithmetic Teacher*, Vol. 22, No. 7, November 1975, pp. 546–554.

Brown, Stephen I. 'From the Golden Rectangle and Fibonacci to pedagogy and problem posing', *Mathematics Teacher*, Vol. 69, No. 3, March 1976, pp. 180–186.

Brown, Stephen I. *Some 'prime' comparisons*. Reston, Virginia: National Council of Teachers of Mathematics, 1978.

Brown, Stephen I. 'Ye shall be known by your generations', *For The Learning of Mathematics*, No. 3, March 1981a, pp. 27–36.

Brown, Stephen I. 'Problem posing: the problem generation gap', *Math Lab Matrix*, No. 16, Fall 1981b, pp. 1–5.

Brown, Stephen I. Distributing Isomorphic Imagery, New York State Mathematics Teachers' Journal **32**(1) Winter, 1982a.

Brown, Stephen I. 'On humanistic alternatives in the practice of teacher education', *Journal of Research and Development in Education*, Vol. 15, No. 4, Summer, 1982b, pp. 1–12.

Brown, Stephen I. and Walter, Marion I. *The art of problem posing*. The Franklin Institute Press, 1983.

Dewey, John. *Reconstruction in philosophy*. Henry Holt and Co., N.Y., 1920.

Dewey, John. *How we think*. Heath, N.Y., 1933.

Getzels, Jacob W. and Jackson, Philip W. *Creativity and intelligence: explorations with gifted students*. John Wiley & Sons, N.Y., 1961.

Gilligan, Carol. *In a different voice*. Harvard University Press, Cambridge, 1982.

Higginson, William C. *Toward mathesis: a paradigm for the development of humanistic mathematics curricula*. Unpublished doctoral dissertation, University of Alberta, 1973.

Hilton, P. Current Trends in Mathematics and Further Trends in Mathematics Education, For the Learning of Mathematics **4**(1), (February 1984) pp 2–8.

Hofstadter, Douglas. 'Metamagical themas: variations on a theme as the essence of imagination', *Scientific American*, October, 1982, pp. 20–29.

Kohlberg, Lawrence. 'Moral stages and motivation: the cognitive developmental approach', in Lickona, Thomas (editor), *Moral development and behavior: theory, research and social issues*. Holt, Rinehart and Winston, 1976, pp. 31–53.

Lickona, Thomas (editor). *Moral development and behavior: theory, research and social issues*. Holt, Rinehart and Winston, 1976.

Lipman, Matthew, Sharp, Ann S., Oscanyan, Frederick. *Philosophy in the classroom*. Institute for the Advancement of Philosophy for Children, New Jersey, 1977.

Lyons, Nona P. 'Two perspectives: on self, relationships and morality', *Harvard Educational Review*, Vol. 53, No. 2, May 1983, pp. 125–145.

McGinty, Robert L. and Meyerson, Lawrence N. 'Problem solving: look beyond the right answer', *Mathematics Teacher*, Vol. 73, No. 7, October 1980, pp. 501–503.

Oliver, Donald and Newman, Fred. *Clarifying public controversy: an approach to teaching social studies*. Little, Brown, 1970.

Snow, C.P. *The two cultures: and a second look*. The New American Library, New York, 1959.

Taba, Hilda. 'The problems in developing critical thinking, *Progressive Education*, 1950, pp. 45–48.

Walter, Marion I. and Brown, Stephen I. 'Problem posing and problem solving: an illustration of their interdependence', *Mathematics Teacher*, Vol. 70, No. 1, Fall 1977, pp. 4–13.

17

Freedom and Girls' Education: A Philosophical Discussion with Particular Reference to Mathematics

ZELDA ISAACSON

> Men do not want solely the obedience of women, they want their sentiments. All men, except the most brutish, desire to have, in the woman most nearly connected with them, not a forced slave but a willing one; not a slave merely, but a favourite. They have therefore put everything in practice to enslave their minds ... The masters of women wanted more than simple obedience, and they turned the whole force of education to effect their purpose.
>
> (John Stuart Mill, The Subjection of Women)

> Well, but now, Casaubon, such deep studies, classics, mathematics, that kind of thing, are too taxing for a woman—too taxing, you know.
>
> (George Eliot, Middlemarch)

INTRODUCTION

Feminists past and present have been much concerned with freedom, with the liberation of girls and women from the constraints and restrictions put on them by a patriarchal system of society. And gradually, over the last century and a half in this country, opportunities for women to take more control over their lives, to choose how to earn their living and to participate in those areas of public life which interest them have increased. Our Victorian sisters would undoubtedly envy us our freedoms. There are women doctors and engineers, lawyers and bus drivers, physicists and plumbers. There are women at every level of government and public service, from school governing bodies to prime minister. Women can choose to marry or not, to have children or not, and to bring up their children in single-parent homes or in communes if they wish. What a contrast with the nineteenth century! Then, universities and professional bodies were closed to us, we did not have the vote—or decent contraception—and our husbands (if we married) owned us, our children and our property. Why, then, should freedom still be an issue? And what has freedom to do with girls' mathematics education? After all, all children study mathematics at school. It is a core subject in the curriculum right through the years of compulsory schooling. No one is preventing girls from studying mathematics—so what is the problem?

This chapter is an attempt to show why I believe that a philosophical analysis of freedom in the context of girls' education can throw light on a phenomenon with which we are all familiar, that is, that despite the advances which have been made, gross inequalities between the sexes remain, in every field of human endeavour and activity. Gross inequalities also remain in income from employment and ownership of property, and indeed in all the ways in which people gain control over their own lives. Opening doors and increasing opportunities has not been sufficient to ensure equality of achievement between women and men, or to redress the imbalance of power which exists in our society. Reasons offered to account for this continuing phenomenon range from the age-old claim of inherent female inferiority, at one extreme, to Marxist-feminist analyses of class (where women play the role of the proletariat for their male masters), at the other. These varied analyses are useful and valuable, I would argue, for I believe that the reasons are so complex that no one analysis can possibly be adequate on its own. We can no more expect to understand what is happening to girls and women in a simple, mono-dimensional way than we could expect to judge the shape of a mountain from one photograph taken from a particular angle. My aim in this chapter is to offer a perspective—just one amongst many possible—on the causes of gender inequality in our society. My vehicle is a philosophical analysis of freedom, for I believe that there is a sense in which girls are less free than boys to take up certain school and career options, and possibly also a sense in which boys are less free than girls to reject such options.

I shall be looking at a range of school subjects and career options with a particular focus on mathematics achievement. There are two reasons why looking at mathematics is likely to be particularly fruitful in this context. Although part of the core curriculum, it is a paradigm case of a school subject which the population as a whole considers 'male' or 'masculine', to which girls and boys often respond differently, and where different achievement levels become apparent long before compulsory participation ends at 16+. The second reason is that mathematics acts, as Lucy Sells pointed out, as a 'critical filter' (Sells 1973) in determining whether an individual can go along many training and career paths. It is noteworthy that a mathematics qualification is a necessary prerequisite for entering many of the areas of employment where gender inequality is particularly marked, such as science, engineering and technology. Also, opportunities for employment within many fields of work are widened by having competence in mathematics. For example, a graphic designer with some knowledge of (and preferably a qualification in) mathematics will have greater scope for employment than an artist without this training, and the same is true in commerce and many other fields. So both in its own right as a school subject which, although studied by all children, nonetheless carries a marked gender bias, and as an essential requirement for entry into many male-dominated areas of work, mathematics is of interest and importance to those who want both to understand the reasons for gender inequality in society and to affect the imbalances we see around us.

FREEDOM AND EDUCATION

The concept of freedom has a long history and is notoriously difficult. Freedom is always context bound and never absolute. To ask whether someone is free, or to be told that they are or are not free is meaningless unless we are given further information. We need to know what they are free from (free from poverty? free from prison?) or what they are free (or not free) to do—or say or think or be. We need to know whether their freedom has time or space constraints, and possibly how much freedom the person has—perhaps in relation to other people.

Consider, as an example, the question, 'Are girls free to choose to study physics as a subject in the fourth year of secondary schooling?' At a very superficial level the answer is 'yes'. There is no law which debars girls from studying physics, and physics is (these days) taught in most schools. Less superficially, a great deal needs to be known before one can formulate an answer. We need to know, at least, which girls we are talking about, and in which schools, whether physics is in fact available as an option, what the timetabling arrangements are, what the prerequisites are for studying physics in the fourth year, whether girls know they may choose physics, and whether the girls—if in the end we conclude they are free—are 'as free as' the boys, and so on. For example, we might decide that the girls in school *X* are not 'as free as' the boys in the same school because physics, although ostensibly equally open to all, is timetabled against child development, a subject which many girls want, and are particularly encouraged, to take. (I am not suggesting by this example that school *X* deliberately conspires against girls in this respect. It is just as likely, perhaps even more likely, that the reason for this practice is that past experience has shown that girls in fact very rarely want to study physics, and frequently wish to do child development, while the reverse is the case for boys. By timetabling these subjects at the same time, fewer pupils are disappointed by being unable to do the subject of their choice because of clashes than would otherwise be the case. On the face of it, the school has provided greater freedom—for pupils to take their first choice subject.) Nevertheless, maintaining this timetabling arrangement reinforces traditional subject choices, not least by making physics less of a viable option for girls than for boys.

An even clearer example is provided by the case of craft, design and technology. Unlike science, where in most secondary schools all children follow the same basic or introductory courses in the first few years, CDT is overtly male dominated much earlier. As Martin Grant, reporting on the GATE project commented '... the pattern for some girls is 'no access' and for most a diminishing involvement in the early years ...' (1983, pp. 20–21). Can girls be said to be free to choose CDT as a fourth- and fifth-year option if they have not previously taken part in foundation courses? This is a difficult question and one to which I shall return later.

It is perhaps necessary to make it clear at this stage that the free-will/determinism debate is not going to form part of my discussion. In discussing types of freedom, degrees of freedom and conditions which need to be satisfied before we can reasonably claim that a person is free (in some sense) to be, do, feel, think something, I shall be taking as an underlying assumption that the concept of freedom has application. That is to say, the notion that we are all actors and agents, and not merely being acted upon by forces wholly outside our control is an assumption in all that follows.

I made a claim, earlier, that in discussions about freedom it is invariably necessary to ask further questions. If someone has said 'I want to be free', one might then ask, 'what do you want to be free to do? (or be? or say?)', or, 'what do you want to be free from?'. However, it is not always possible to answer such questions. A person may feel constrained but not know why, or perhaps not know what s/he would do if not so constrained. For example, a woman with young children at home might feel trapped by her situation yet have no particular ideas of what she might do if she were free of these ties. As Isaiah Berlin says:

> 'A man struggling against his chains, or a people against enslavement need not consciously aim at any definite further state. A man need not know how he will use his freedom; he just wants to remove the yoke.'
>
> (Berlin 1969, p. xliii)

Having said this, it is nonetheless true that an awareness of the chains, the enslavement, the restrictions on one's possible actions, the burdens or miseries of one's situation, are an essential part of the desire for freedom from those constraints. Lack of such awareness may not preclude others from claiming that we ought to want release from our chains, but it will preclude us from wanting it ourselves. This is an important consideration in the context of the education of girls, many of whom could be argued (by educators and feminists) to be constrained in their choices of school subjects and future careers even if they themselves are wholly unaware of these constraints, and believe they are making 'real' and 'free' choices. To bring the discussion around to mathematics, most girls who 'vote with their feet' in mathematics classes, and end up in CSE instead of 'O' level groups, and with poorer qualifications than they are capable of getting, would be indignant if anyone suggested that their freedom to study mathematics had been curtailed. They would claim that they were not interested in mathematics and had chosen not to make an effort in it—the 'I can't be bothered with maths, it's no use to me' kind of response—without any realisation that the fact that the belief (false in many cases) that mathematics is of no use to them could itself constitute a restriction on their freedom. In the same way, girls who 'choose' child development rather than physics, and cookery rather than CDT, generally believe they are making free choices.

It is for these sorts of reasons that putting on events such as the 'Be a Sumbody' Conferences (see Chapter 15) which are designed with the twin aims of showing girls some of the enjoyable aspects of mathematics and a range of career options for which mathematics is needed, and the Insight programme (Peocock and Shinkins 1983) which aims to inform girls about engineering careers are so valuable. Information and interest are necessary components for creating the desire which is an important element in the exercise of freedom.

Why is it that these girls (and often their parents and teachers) believe that they are freely choosing, e.g. cookery, while some people question this? In part this might arise from the fact that while there is a tendency in our society to think of freedom in terms of the absence of legal or other formal constraints, this is regarded by others as too narrow a view of freedom. A consideration of negative and positive freedom and of the elements involved in the exercise of freedom may help to clarify this. The distinction between negative and positive freedom has been particularly well drawn by Isaiah Berlin (1969) and I am indebted to him for my understanding of this.

Negative freedom and positive freedom

Negative freedom is the absence, or the reduction to a minimum, of the deliberate interference of others in the areas in which I wish or might wish to act. Those who opposed the introduction of a law compelling motorists to wear seat belts were making a claim for negative freedom. If such a law does not exist, I am free from a legal constraint on my actions, free to choose whether or not to wear a seat belt. Negative freedom thus involves an absence of the sorts of constraint which are, in principle at least, apparent to all. That there are these constraints on my freedom remains true even if in fact I do not wish to do the things I am not permitted to do. And the reverse is also true. If I am a teetotaller (or a vegetarian) I am nonetheless free (in the negative sense of freedom) to have an alcoholic drink (or a steak) if I choose to. The fact that few girls in England study physics, engineering and technical subjects does not alter the fact that many more are free (again in a negative sense) to do so.

The existence of negative freedom is not, however, sufficient to ensure that justice is done or even that people can exercise this freedom. Berlin comments on:

> ... a situation in which the enjoyment by the poor and weak of legal rights to spend their money as they pleased or to choose the education they wanted ... became an odious mockery. ... Legal liberties are compatible with extremes of exploitation, brutality and injustice.
>
> (Berlin 1969, p. xlvi)

Berlin's insight applies as much to contemporary social and educational situations, where an uncritical application of the term 'freedom' can be used to conceal aspects of education and society about which we ought not to be complacent. Thus, in the case of school *X* discussed above, emphasis on the girls' formal right to choose physics conceals the injustice of the actual situation for many of them, who would perhaps have chosen physics, an immensely useful subject in terms of future careers, had they been advised of its importance for them and had it not been timetabled against child development.

This kind of injustice is often blatant in the case of technical subjects. In many junior schools girls and boys are introduced to different practical subjects. As the DES survey on curricular differences for boys and girls (1975) commented:

> ... timetabling arrangements may provide for girls to learn the skills of needlecraft, and for boys to engage in a wider variety of crafts involving the use of a range of tools and materials leading to three-dimensional modelling and construction and the use of measurement.
>
> (DES 1975, p.3)

Although ten years have passed since this survey was carried out, there are still huge gender differences between girls' and boys' school experience of craft by age 11, and only one in eight secondary schools surveyed by GATE 'could claim equal representation of girls and boys throughout the first three years' (Grant 1983). In most secondary schools the differentiation begun earlier is continued, with girls cooking and sewing while boys do woodwork and metalwork. These curriculum differences do a great deal of damage, not least in that they lead to different achievement levels in cognitive skills useful for learners of mathematics and science. They also, of course, serve to reinforce the sexual stereotyping about appropriate activities for children (and adults) which are so solidly established in our society.

Whereas negative freedom can be conceived of as the answer to the question 'In what areas of my life may I act without interference from others?' positive freedom is to do with self-realisation, with 'being my own person', in other words as an answer to the question 'by whom am I ruled?' The desire for positive freedom arises from the desire of the individual to be in charge of her own life. None of us can live without restraint of any sort. We can, however, to a greater or lesser extent choose our own constraints, and the extent to which we make these choices is the extent to which we are self-governed rather than ruled by others. Positive and negative liberty are not the same thing. Indeed, they may clash. If girls have negative freedom to study physics, equally they have negative freedom to choose not to study this. It could be argued that girls' negative freedoms—to choose whether to study physics and technical subjects, or whether to apply themselves to their mathematics—are at time in conflict with the need for girls to study these subjects if they are to have positive freedom, that is, to be in charge of their own lives, as adults. This is the standard justification for compulsion in education to which I shall return later.

Elements involved in the exercise of freedom

That freedom and the conditions necessary for the exercise of freedom are not the same thing has already been asserted, as has the injustice of situations where a legal right exists but people are not able to exercise it. There are a number of distinct elements involved in the exercise of freedom. These are:

(a) whether preconditions necessary for the exercise of the freedom have been satisfied;
(b) whether internal constraints exist (e.g. psychological constraints);
(c) whether the desire to exercise the freedom is present;
(d) whether the person concerned knows that the freedom exists;
(e) whether the person has the capacity to exercise the freedom (e.g. intellectual, physiological).

Particular cases in which a (negative) freedom exists but is not exercised may be due to any one (or a combination) of these. And debate is often confused because of a widespread failure to separate them. Some examples may serve to clarify this.

It is well known that very few women take up engineering as a career, despite the fact that there are now no formal barriers to this. The reason for any particular woman not making this choice could be any of the elements outlined above. I shall invent hypothetical (but typical) women and ask them why they didn't choose engineering. Here are their replies:

Peggy: 'I thought about taking up engineering when I was leaving school, but I hadn't done physics so I didn't have the right qualifications. I wasn't willing to spend even more years doing school exams.' (Case (a)—necessary preconditions not satisfied.)
Lesley: 'I was interested in engineering, but everyone I talked to told me how much of a mucky man's world it was, with women made to feel uncomfortable. I didn't think I could cope with that.' (Case (b)—existence of internal constraints.)
Marion: 'I could have taken up engineering, but it really didn't grab me. I'm more interested in pure science and I'm doing a chemistry degree now.' (Case (c)—desire absent.)

Kelly: 'What a funny question—engineering's for men, isn't it? Women can't do that!' (Case (d)—knowledge that women can (and do) become engineers is absent.)
Frances: 'I'd have liked to have done engineering, but it's a very active job which needs good health, and since my accident I'm really not fit enough to do that sort of work.' (Case (e)—capacity inadequate in some respect.)

This is, of course, a gross over-simplification. Why hadn't Peggy done physics at school? Marion seems to have made a positive choice, but what about other women who 'don't have the desire' to study science or technology? This is not always because they have made positive choices in a different direction. If Lesley's analysis is accurate, then she is being prevented from taking up engineering for reasons extraneous to her capacity and desire to do so. Similarly, a woman who 'chooses' a less demanding, perhaps part-time, job which enables her to care for her children properly, and does not contemplate a full-time 'man's job' because of the lack of adequate child-care facilities, is being prevented by reasons other than capacity or desire. Is this an absence of freedom in a real sense?

THE SEX DISCRIMINATION ACT

In this section I shall look at the educational provisions of this Act and indicate why I believe it has proved an inadequate tool in establishing freedoms and combating long-established inequalities.

Even a cursory reading of the educational clauses of the Sex Discrimination Act is sufficient to establish that this gives girls and women only formal freedoms. Only external constraints are eliminated, and then not all of these. The Act states that it is unlawful for the responsible body of an educational establishment to discriminate against a woman:

> (a) in the terms on which it offers to admit her to the establishment as a pupil; or
> (b) by refusing or deliberating omitting to accept an application for her admission to the establishment as a pupil; or
> (c) where she is a pupil of the establishment:
> (i) in the way it affords her access to any benefits, facilities or services, or by refusing or deliberating omitting to afford her access to them, or
> (ii) by excluding her from the establishment or subjecting her to any other detriment.
>
> (Sex Discrimination Act 1975, extract from section 4.2)

It is unclear what 'deliberating omitting' in (c)(i) means. Is there any way to judge when 'omitting' becomes 'deliberating omitting'?

So it is unlawful for a school which admits girls and boys to refuse a girl admission to the school or to a physics or CDT class on the grounds of her sex alone, and it is unlawful for further and higher education establishments to refuse a girl admission, say, to an engineering or a building trades course on the grounds of her sex. However, there is no requirement on anyone to inform girls that these courses are available to them and likely to be useful to them, or in any way to attempt to break down the mythology surrounding these subjects and career choices. Schools need not even refrain from actively encouraging girls to take subjects and aim for careers deemed more appropriate for them. Nor, of course, is any school obliged to provide, e.g. physics, computing or CDT in the first place, and the importance of this becomes clearer when we reflect on the fact that girls' schools seldom have technical facilities and that science laboratories are rarer in girls' than in boys' schools. Nor does the law require child-care facilities to be provided for working mothers, a factor which greatly influences adolescent girls' choice of careers. On the contrary, such facilities are regarded by many as a dispensable luxury which indeed we have seen being rapidly depleted as unemployment has grown and women have again been told that their place is in the home.

The reality of adolescent 'option choice' time in most secondary schools is that by then not only have both girls and boys thoroughly absorbed society's messages about the gender-appropriateness of different school subjects and future careers and jobs, but also the school's timetabling arrangements and the guidance offered to pupils will reinforce these messages. To add additional weight to these messages, staff teaching the physical sciences and technical subjects may well advise pupils not to take these subjects if they have not already reached an appropriate minimum standard in them, as this is deemed necessary if they are to be taken to examination level in the fifth year. This ruling eliminates some of the boys and many more of the girls, especially in technical subjects, where girls, as discussed above, have had far less experience by this stage of their schooling. Thus while the CDT course, say, may be ostensibly open to all pupils, a covert sex-bar is in fact being operated. This is not acknowledged, however, as the school typically characterises the bar as a (legitimate) educational one, and characterises its activity as one of 'advice' rather than legislation.

Should parents, indignant on behalf of their daughter, and aware of the injustice, invoke the Sex Discrimination Act, they would be told that no discrimination on the grounds of sex is being operated and that if they insist, their daughter may join the class. Those 'advising' may well add that in that case the school would take no responsibility for her success in the course as this decision runs counter to their considered advice. This is indeed an 'odious mockery' parading under the name of free choice.

Thus the Sex Discrimination Act changed the facts about overt discrimination and gave many girls more negative freedoms than they had previously. But the Act did nothing, in itself, to ensure that girls become aware of those changes that there are in the range of options open to them. And it did nothing to break down existing prejudices and firmly held beliefs which, together with objective external circumstances outside the jurisdiction of the Act, militate against the exercise of these choices. Removing formal constraints, although necessary, is far from sufficient to ensure that a freedom (rather than a mockery of a freedom) exists. A freedom, to be of any value to anyone, must be *de facto* utilisable, not merely *de jure* in existence.

MATHEMATICS

All pupils study mathematics from the very outset of formal schooling and continue to do so until the end of the compulsory phase of education. Nonetheless, this situation of apparent equity conceals important differences. Differences of outcome, in terms of 'O' level passes and grades are well documented and do not need reiteration (e.g. Isaacson 1982). Being in a mathematics class is not enough to ensure achievement. What matters is both which mathematics class and what you do once you're in it. The gradual downward drift of girls from top sets to lower ones as they go through secondary schooling is all too familiar. When I took a top maths set from the second year through to 'O' level a few years ago, in a London comprehensive school, the proportion of girls in this class altered from two thirds in the second year to one third by the fifth year. Equally well documented is poor female participation at 'A' level mathematics and beyond. And the good proportion of girls getting a CSE qualification in mathematics is not really cause for celebration, as many of them could and should have taken 'O' level, thus improving the 'O' level statistics at the expense of CSE.

Girls' excellent achievement in mathematics at the primary stage and through to the lower forms of secondary school is not maintained. In Chapter 5 of this volume Rosalinde Scott-Hodgetts argues that the way mathematics is taught at primary level is disadvantageous, in the long run, to girls. While this may in part account for girls' gradual decline in achievement, it is not adequate to account for the scale of the decline. Of at least equal importance are other factors—factors related to the preceding discussion.

Mathematics is not only a subject in its own right, it is also a service subject. If you want to study physics or engineering, architecture or accountancy, chemistry or computer science—all subjects and careers with a 'male' image—you will need mathematics. So boys, who are encouraged to consider their careers important, and are guided towards these kinds of choices, are also encouraged, explicitly and implicitly, to work at their mathematics.

At option choice time, boys are guided towards the physical sciences and technical subjects. Pupils, girls and boys, who study these subjects tend to do better at mathematics than those who do not, which is hardly surprising as a sizeable proportion of the curriculum content builds on and develops mathematical knowledge and understanding. The research carried out by Sharma and Meighan points in this direction (1980, pp. 193–205). It could be argued, of course, that pupils who are 'no good at maths' don't choose these subjects in the first place because of their weakness in mathematics, but experience does not bear this out. In some cases this is indeed the direction of causality, but for many girls the decision not to do sciences or technical subjects at 14 is linked to the masculine image of the subjects themselves and of the jobs they lead to. The fact that girls more readily 'choose' these subjects in single-sex than in mixed schools is significant. In a situation where a girl's gender identity is under threat, and this is potentially greater in a mixed than a single-sex school, she is more likely to reject any work identity which further threatens her. She'll play safe, in other words, and choose 'girl' subjects and career aspirations.

The effect on mathematics achievement is thus cumulative. Mathematics itself has a male image—and television programmes like the BBC *Horizon* programme *A Mathematical Mystery Tour* broadcast in December, 1984 do nothing to alter this—so many girls will not try to succeed in it at a stage when being overtly 'feminine' is particularly important to them. Girls study physical science and technical subjects, for this and other reasons, less often than boys, so the reinforcement these can give to mathematical skills and the motivation provided by them for working at mathematics is lacking for them. Girls seldom have or are encouraged to have career aspirations in directions where mathematics is clearly going to be needed, so again their motivation to work at mathematics is lower than that of the majority of boys. Although these girls are physically in mathematics classes their hearts and spirits are not there, and this is reflected in their performance.

Conventional school practices, as discussed above (such as timetabling arrangements, careers advice, and guidance of pupils towards 'gender-appropriate' activities) all reinforce the status quo and make it more difficult for young people to exercise the negative freedoms open to them. And the outcome for girls is under-representation in science and technology, and, in mathematics, under-achievement by age 16, and under-representation thereafter.

COERCIVE INDUCEMENTS AND COMPULSION, DOUBLE BINDS AND DOUBLE CONFORMITY

The notion of a coercive inducement, which, as the name suggests, is an offer which can't be refused, offers a further perspective. There is a sense, I believe, in which many girls are persuaded to adopt traditionally female modes and behaviour and to choose stereotypically feminine occupations and life-styles because the rewards for 'feminine' behaviour are too great to be refused, rather than because they are prevented from choosing others. Early in the girl's life the kinds of rewards I have in mind are approval (isn't she a helpful little girl) so that the six-year-old 'chooses' to help mother with the chores or teacher with the tidying up instead of going out to climb trees or kick a football in the playground. In the long term, such 'helpfulness' is not in the girl's own best interests as she is missing out on developing not only spatial awareness and skills but also physical confidence and independence, all of which are fostered by these typical boys' activities. Later on, the rewards for her growing femininity are being thought to be attractive, being taken out on dates and treated by boyfriends, and eventually, the ultimate goal of so many girls, marriage and motherhood. These are rewards which our society makes girls value so much that there is a sense in which the vast majority of girls can't refuse them. It is arguable whether the typical girl who chooses acceptably feminine school subjects and activities, whose chief aspiration is marriage, and who spends most of her time and energy thinking about boys and how she will appear in their eyes—hence the obsession of so many teenage girls with clothes and makeup—is making free choices. She would claim that she is, and that this is what she really wants, that a job is just a way of filling in time and earning some money (to buy more clothes and makeup) before marriage, and therefore not worth spending much energy on. The reality, I would argue, is, however, that many girls are being coerced by an inducement they can't refuse. Both coercion and seduction are at work, and if girls don't get the message in other ways they are further controlled by boys' abusive language. All girls, unless they 'belong' to some male (i.e. are going steady, are engaged or married) run the risk of being labelled as 'slags' or 'tight bitches'. The only way to escape such abuse is by playing things the boys' way. So girls are told, loud and clear, how wonderful it will be to be a bride, to be married, to have babies, and what failures they will be if they don't achieve these things. Little wonder that the desire to work at difficult, and 'non-feminine' school subjects is so frequently lacking, and that girls' career aspirations are so often disappointingly low.

When Berlin stated that

> to be free—negatively—is not simply not to be prevented by other persons from doing whatever one wishes ... If degrees of freedom were a function of the satisfaction of desires, I could increase freedom as effectively by eliminating desires as by satisfying them.
>
> (Berlin 1969, p. xxxviii)

he could have been (although of course he wasn't) talking about girls' education. For by very largely creating in girls the kinds of desires which will lead to conventional life choices, and not creating in them those desires—for a fuller and freer (in the sense of positive freedom) life—which society finds uncomfortable in women, the way is left open for the claim to be made that women have these freedoms but don't want to use them. The situation of girls is even worse than that alluded to by Berlin in that not only have some desires been eliminated but as well other desires have never even been awakened.

Many people say that the way to awaken girls' interest in science and technology is to make these subjects compulsory. Undoubtedly, all pupils ought to be introduced to the elements of science and technical subjects in the primary school and craft 'circuses' in the lower secondary school should be organised so as to give a genuine opportunity for all pupils to get a taste of all the crafts and learn the basic skills. As has already been pointed out, girls don't even get the chance to 'choose' CDT at 14 if they've had no introductory courses and lack the skills deemed necessary as a foundation. But compulsion alone will not solve the problem and itself creates other problems. Unless girls have the desire to study these subjects, by adolescence they will 'vote with their feet' in their physics and CDT classes, and drop down (just as they already do in mathematics) if they are not allowed to drop out. We have to change the way we present these subjects to girls, and also change the images we give them of the subjects and of themselves as future adults. It would be morally preferable to make it easier for girls to exercise their existing negative freedoms (and thus pave the way to greater positive freedom as adults) rather than reduce their negative freedoms now. Changes of this far reaching sort are extremely difficult to bring about, however, and take a long time. In the short term, we may decide that compulsion is unavoidable, and that what we ought to do is restrict girls' freedom now in order to give them greater freedom later, in a defence exactly analogous to that for compulsory education.

In addition to the moral, there is a practical problem. Pupils who are forced to study subjects they don't want to do make unwilling and often rebellious students, who frequently become even more fixed in their aversion to the subject. So it is open to argument whether more good or more harm would be done by forcing girls (or boys) to study subjects they would rather drop. We see this very clearly in mathematics, where as already discussed many pupils, especially girls, drop down (in level of attainment, school set, and examination taken) because they are not permitted to drop out.

So it seems that a situation exists in which the only way many people see of achieving equality of outcome for girls or of paving the way to their greater positive freedom as adult women is by removing some of the existing freedoms of the girls they are trying to emancipate, through forcing them to study subjects they do not wish to take. This is a double bind, and we would do well to remember that, as Berlin puts it:

> ... negative and positive liberty are not the same thing. Both are ends in themselves. These ends may clash irreconcilably... If the claims of two (or more than two) types of liberty prove incompatible in a particular case, and if this is an instance of the clash of values at once absolute and incommensurable, it is better to face this intellectually uncomfortable fact than to ignore it ...
>
> (Berlin 1969, pp. xlix–l)

I would suggest that this is such a case, and that we might get further in our discussions if we were more willing to recognise 'this intellectually uncomfortable fact'.

Another kind of double bind is that created by the demands of double conformity, a phenomenon which all women who have stepped out of stereotypically female roles and occupations have experienced. I have discussed in some detail the way in which coercive inducements affect girls' and women's life 'choices'. A consequence of coercive inducements is that the majority of women make choices which result in behaviours which conform very closely to society's expectations of them. However, even women who reject these stereotypical choices are still expected to conform to the general behaviour patterns which society has deemed 'feminine'. In addition, these women have to conform, in their studies and their work, to the 'correct' or 'appropriate' ways of doing these 'masculine' subjects or jobs. As these have been defined by men for men, conforming to these norms is often in conflict with 'feminine' behaviour or ways of thought. It is much harder for a woman to be successful in these subjects and careers than it would be for a man. She not only has to have the necessary intellectual and practical capacities, she also has to be secure enough in herself to make rational decisions about appropriate behaviour in various circumstances and withstand mockery and put-downs where her behaviour fails to conform in one way or another. This it inevitably will do, as consistent conformity is impossible.

One way out of this particular double bind is to redefine 'femininity' and another is to redefine what counts as appropriate behaviour in these jobs. Even more far reaching, and potentially very significant, are recent attempts to redefine the nature of the subjects concerned. Some of the most interesting developments in science education particularly, and more recently in technology and mathematics, have been concerned so to redefine the content, context and methodology of these subjects as to make it possible for women to engage in them without these sorts of conflict. As well, a growing number of people are concerned at the dangers to our society from 'male' science, technology and mathematics. This arises from a recognition that the rigidity, lack of concern with ethical questions and lack of human relatedness which characterise these disciplines as they are generally practised today, are taking us in increasingly ominous directions. To humanise these disciplines would be to incorporate typically 'female' concerns, modes of behaviour, insights and responses and this could be considered necessary for a saner safer world. Although not intended primarily to encourage more women to participate in these disciplines, such changes would undoubtedly have this effect. Of particular interest in this context is Brian Easlea's work which is concerned mainly with science (Easlea 1981, 1983). The need to humanise mathematics is discussed by Stephen Brown in Chapter 16 of this volume.

One could go further and ask whether boys have been persuaded (coercively induced?) to participate in science, mathematics and technology to a far greater degree than they 'really' want to. Sheila Russell studied sixth-form pupils in Bradford, and found that although, as nationally, there were far fewer girls than boys taking 'A' level mathematics, most of the girls who were doing mathematics had chosen it because they liked it, and were enjoying their sixth form course. Many of the boys, on the other hand, were not enjoying the course, and had 'chosen' mathematics because this was expected of them or because it would be useful in the future (Russell 1983).

It may be the case that more boys than girls achieve well in mathematics because they are less free than the girls to reject it and pursue other studies rather than because girls are less free than boys to choose it! The truth probably lies somewhere between these extremes. If there were fewer gender-stereotyped expectations laid on young people, more of them would be free to choose options outside the gender norms, with the result that more girls and fewer boys would choose science, mathematics and technical subjects. Changing the ways these subjects are both defined and perceived would also in the long-term result in them being seen as human rather than male, and this too would lead towards greater gender equity in participation in them, which would be good for women, good for the subjects themselves and good for humanity.

If these changes can be achieved then freedom and equality in girls' education become realistic goals. Until such time—and fundamental changes like these take a long time—we probably have to accept the uncomfortable fact that freedom and equality are irreconcilable in girls' education, and that different types of freedom are also irreconcilable, and make decisions based on value judgements in each situation.

My conclusions are therefore that we need to work slowly, in an evolutionary way, towards a new, more humane vision of mathematics, science and technology and at the same time also work at breaking down unnecessary gender stereotypes in behaviour and career aspirations. A more revolutionary approach would, I fear, result in losses we can ill afford—of existing freedoms, and of valuable 'female' modes of thinking and behaviour.

REFERENCES

Berlin, I. (1969) *Four Essays on Liberty*. Oxford: Oxford University Press.

DES (1975) *Curricular Differences for Boys and Girls*, Education Survey 21. London: HMSO.

Easlea, B. (1981) *Science and Sexual Oppression*. London: Weidenfeld and Nicolson.

Easlea, B. (1983) *Fathering the Unthinkable*. London: Pluto Press.

Grant, M. (1983) 'Mathematics counts in craft, design and technology—but not for some' in *GAMMA Newsletter*, **4**, pp. 20–21.

Isaacson, Z. (1982) 'Gender and mathematics in England and Wales' in *International Review of Gender and Mathematics*, ERIC.

Peocock, S. and Shinkins, S. (1983) Review of the INSIGHT programme designed to encourage girls to become engineers, Engineering Industry Board.

Russell, S. (1983) *Factors Influencing the Choice of Advanced Level Mathematics by Boys and Girls*, Centre for Studies in Science Education, University of Leeds.

Sells, L. (1973) 'High school mathematics as the critical filter in the job market' in *Developing Opportunities for Minorities in Graduate Education*. Proceedings of the Conference on Minority Graduate Education at the University of California, Berkeley, May.

Sharma, S. and Meighan, R. (1980) 'Schooling and sex roles: the case of GCE 'O' level mathematics', *British Journal of Sociology of Education,* **1** (2).

18

Girls and Technology

ROBIN WARD

BACKGROUND

For a number of years there has been widespread concern about sex-related learning differences in science and mathematics, and it is not surprising to find these differences emerging in the area of computing. Because computers tend to be associated in most people's minds with mathematics, girls who do not like maths or who feel they fail in it, will tend to transfer this sense of failure to using the computer and learning about information technology. The pattern is similar to that observed in science (EOC 1982). An area of study becomes associated with male dominance and all the images and behaviours related to it reinforce that association. For example, the traditional sex-stereotyping and male domination of texts and media, presently being tackled in relation to maths, has appeared in all areas of computing; a full page of a well-known computer magazine asks 'Why not buy your son a computer for Christmas?' At least nine times as many boys as girls are likely to have a computer in their homes and manufacturers of both hardware and software consciously direct their advertising at the male consumer. Another blatant example of male-dominated advertising suggests that anyone buying a computer could ask 'father or elder brother' for any help that might be needed.

Posters advertising jobs in science, engineering and technology tend to portray the man in the dominant, responsible post while the woman, if she appears at all, waits with pencil poised, or fingers hovering over the typewriter (computer) keys, for the words of her lord and master. All these examples, and many more, help to perpetuate the divisions between males and females.

Aspects of girl/boy differences in learning style, such as the dominance of aggressive boys in class, which are apparent in other areas of the curriculum, have shown themselves also in computing. Many girls do not demand the time and attention they need; fearing ridicule, girls do not volunteer responses in class. And when things go wrong, girls often blame themselves while boys blame the teacher or equipment. Computer rooms, available before and after school, quickly fill with boys, and girls are intimidated enough not to enter. Lack of awareness by many teachers of the effects on girls of a male-biased curriculum, such as tendency to ignore them when the more confident boys demand attention, is one factor which explains girls' alienation. In many schools the option schemes help to steer girls into traditional areas by blocking computer studies against such choices as French, a subject enjoyed by many girls.

It is not difficult for teachers to persuade boys that technology is for them; the problem is to convince girls that they too could benefit. When computer studies was first introduced into schools, many girls were discouraged from choosing the subject as an option because of its implied link with the traditionally male mathematics. The computer equipment was stored in the 'maths block' and many of the teachers prepared to tackle the new subject were mathematicians, usually male.

In a survey carried out in Croydon in 1983, 48 per cent of computer studies teachers were found in the mathematics department (the rest were from specialist CS departments) and 75 per cent of all computer studies teachers were male (London Borough of Croydon 1983).

In England as a whole, girls formed 29.2 per cent of the total entries for computer studies in 1980 (DES 1980). Of the 324 Croydon pupils starting a computer studies examination course, completed in the summer of 1982, only 24 per cent were girls. Far too many, 38.8 per cent, of those girls failed to sit the final examination. Some had chosen the subject without knowledge of what the course entailed, were disappointed with the technical aspects and methods of teaching, fell behind and dropped out of the course. Others, when asked in a questionnaire what they might like changed, commented, 'I wish there were more women in the department. It makes me think that computers aren't really for us'. Many of the problems in computer studies were mathematically based such as introductory programming exercises on currency conversion, finding the sum of input numbers or finding an average mark, and this involved anxiety. Recently, a shift has taken place away from courses in computer studies towards courses in computer appreciation or computing through information technology which are seen as being more relevant and interesting.

In the spring of 1983, worried by the small number of girls sitting computer related examinations, the London Borough of Croydon agreed to produce a document entitled *Information Technology in Schools* for the Equal Opportunities Commission. This was to be included in a package which the EOC intended to send to all secondary schools throughout the country as part of the WISE (Women into Science and Engineering) scheme for 1984. Also, it was decided to introduce a course for pupils in the 11–14 age range specifically designed to interest both sexes, to give all students the opportunity to gain an awareness of and experience with computers, and to gain in confidence and competence without the added burden of an external examination. Called 'information technology' the syllabus was created by Croydon in conjunction with the Microelectronics in Education Programme (MEP) and the Department of Trade and Industry.

INFORMATION TECHNOLOGY IN SCHOOLS

The course is made up of a series of twelve pupils' books with supporting software (for BBC and RML machines) and comprehensive teachers' notes. Each of the twelve books provides a complete unit of work on a topic in the syllabus such as information skills, the IT family, and applications and implications. The series is not sequential and, although one volume is an introduction to the course and some units are aimed at the top of the age range, the course can be ordered to suit the strengths and preferences of the teachers involved.

The syllabus was planned to give pupils a wide experience of computers and to show them applications in as many areas as possible. For many, it will be enough to have had the experience of handling and becoming familiar with computers in information technology classes; others may wish to continue their studies towards 'A' level computer science.

Because information technology at 11–14 is not an examinable subject, Croydon has developed a Certificate of Competence to monitor progress. It can be useful to show levels of competency, helps in determining previous experience and can be used as an entrance requirement for extra-curricular clubs. Pupils proceed at a pace which suits them so girls, finding they do just as well as the boys, have a welcome boost to their confidence. The certificate is not intended to be used as an examination and should not be used in a testing environment. The teacher should be able to check off on a check list many of the points from seeing the pupils work in normal lesson time or in a club. There are four levels in the certificate. Each covers a different aspect of computer knowledge and allows the students to gain confidence by rewarding positive achievement over a wide range of computer skills.

The progress of pupils in Croydon as the certificate is introduced has been closely monitored for the first six months. The numbers of boys and girls attaining the various levels has been noted and guidance given to teachers if both sexes do not appear to be achieving equal success.

Although, ideally, information technology is cross-curricular and should naturally be incorporated into a wide range of subjects, IT, at least for the present, is being introduced in Croydon schools as a separate subject. The course is being taught by a variety of teachers from many backgrounds. It is important that teachers who have communication skills and a general interest in technology, rather than in the devices themselves, should introduce the new subject. In the future it is hoped that computer literate teachers, of both sexes, in a wide variety of subjects, will use the technology as a natural feature of their teaching.

However, it was recognised that it was not enough to expect girls to accept information technology just because 'everyone' takes it; after all, 'everyone' takes mathematics! The teachers involved in writing 'Information Technology' identified a number of problem areas for equal participation and suggested possible strategies to encourage it.

Problem area

Texts, other resource material such as posters, videos, articles.

Strategies

Check all teaching materials and other resources for general appeal. If possible, focus on non-traditional roles and avoid stereotypes. (The Equal Opportunities Commission is a useful source and the ILEA Media Resources Centre has produced a very interesting and helpful pack on 'Women in Engineering' which includes a video, posters and teaching material. They are working on a pack on 'Women in Computing' at the moment).

Problem area

Sexist illustrations.

Strategies

Think carefully about the examples used to illustrate a point. Ensure the applications appeal to both sexes equally. Challenge the assumptions in a sexist illustration.

Problem area

Classroom domination by boys.

Strategies

Watch discussions carefully and, if necessary, direct them to avoid dominance by either sex. It is often easy for a teacher to allow a spirited and interesting discussion to carry on without realising that mainly boys are taking part.

Allow girls to group together for moral support. Girls prefer to work together in small groups, pooling their knowledge and coming to conclusions together.

During extra-curricular activities, try to ensure an even spread of the sexes. If permitted, boys will take over. It may, initially, be necessary to institute girls-only days for the computer club. The need should eventually disappear as girls gain confidence.

As far as possible practise a small group discussion pattern in preference to a whole-class model as this forestalls domination by one individual or group.

In addition, schools were encouraged to:

(a) develop a policy of positive action to ensure girls are not discriminated against;
(b) develop an option scheme which would not make the choice of technological subjects difficult for girls;
(c) supervise clubs to stop aggressive boys taking a disproportionate share of available resources;
(d) ensure there is no undue linking of computer technology with what are seen as boys' activities by the choice of staff, location or subject matter;
(e) develop a co-ordinated policy for the identification and training of suitable teachers.

Local Education Authorities, too, were recommended to:

(a) provide sufficient properly equipped rooms;
(b) devise a coherent and comprehensive programme to attract and train the right teachers;
(c) organise specific seminars for head teachers to implement curriculum change;
(d) organise resource centres where good materials are easily available;
(e) ask schools to provide statistics on the relative popularity of different subjects with boys and girls and the relative success in examinations.

EVALUATION

As part of the joint Croydon EOC project, a survey testing attitudes towards computers and computer personnel was given to over 400 Croydon pupils aged from 10 to 16, with the objective of assessing the effectiveness, in terms of the girls' responses, of the IT syllabus and the strategies adopted by teachers and schools when implementing it. The results have not yet been finalised but preliminary findings suggest that the hardest sexist attitudes are held by boys in the fourth year of the junior school. Their replies tend to imply that they see women as second-class citizens in technology and see no reason why they should stray into traditional male preserves. Many women teachers (and most primary teachers are women) take an unduly pessimistic view of their own ability to adapt to technology, particularly if they are not comfortable with mathematics, and this attitude is transmitted to the pupils.

It is interesting to note that even teachers who are totally committed to equal opportunities and who are very aware of the problems faced by girls in technology, still cannot totally avoid boys' dominating the classroom. The report, *Sexism in Schools*, by the Association of Educational Psychologists (AEP), refers to the fact that 'even when teachers were trying to compensate the girls, they gave them less than 40% of their time yet this was perceived by all pupils in the classes as giving more time to the girls than to the boys'. (AEP 1985, p. 4).

A cross-section of pupils hold stereotyped views which firmly place women in inferior positions with regard to careers and prefer to see them at home looking after the children. A female teacher in a junior school in Croydon, dedicated to eradicating stereotypes in the classroom, was horrified to discover the views admitted to in the survey by her male pupils. Despite a spirited class discussion, led by the girls, which ended with the majority of boys agreeing that perhaps girls have the ability and the right to compete in male areas, one boy steadfastly refused to shift to the majority view. However, by bringing it out into the open and holding frank and honest discussions with pupils, teachers and parents, a climate of opinion is established which helps to place prejudicial and stereotyped positions outside the norm.

THE FUTURE

Employment opportunities are becoming fewer for those without skills; many of the new skills are technology based; the jobs available are changing; many of the jobs which are disappearing are those which have been traditionally female. In addition, technology is playing an increasingly important part in most people's lives. What will happen to the girls who leave school in the future if they have been allowed to ignore technological subjects and to develop feelings of inadequacy and incompetency in relation to technology? Confidence and skill in using and evaluating the use of computers is an integral part of the education of future citizens, 50 per cent of whom are women.

REFERENCES AND FURTHER READING

Association of Educational Psychologists (1985) *Sexism in Schools*, AEP, 3 Sunderland Road, Durham, DHI 2LH.

Department of Education and Science (1980) Statistics of School Leavers' CSE/GCE Results England.

Equal Opportunities Commission (1982) 'Girls Into Science and Technology', Research Bulletin **6**.

Hawkins, J. (1984) 'Computers and Girls: Rethinking the Issues', Technical Report **24**, Bank Street College of Education, New York, USA.

Lockheed, M. and Frakt, S.B. (1984) *Women and the New Technologies*, Educational Testing Service, Division of Educational Policy Research and Services, Princeton, NJ, USA.

London Borough of Croydon (1983) *Information Technology in Schools*, Equal Opportunities Commission, Overseas House, Quay Street, Manchester M3 3HN.

Millman, V. (1984) *Teaching Technology to Girls*. Elm Bark Teachers Centre, Mile Lane, Coventry.

Pratt, J., Bloomfield, J., and Seale, C. (1984) *Option Choice—A Question of Equal Opportunity*. Windsor: NFER Nelson

Women's National Commission (1984) *The Other Half of Our Future*. London: Cabinet Office.

Appendix

*Women in Mathematics—Herstory**

History is a dicey business. We tend to think that history books tell us about the best and most important things from the past. But at the same time we know that history books tell us about war, not peace; kings, not farmers; extraordinary events, not everyday life; men, not women. It's hard to know where we stand in relation to history.

The women whom you will read about here are extraordinary women, for two reasons. Firstly, they are extraordinary because they had the chance to learn, and because they had the inspiration and the determination to use their learning. And secondly, they are extraordinary because other people wrote about them, so that we still know about them today.

But many extraordinary women were not written about, and have been forgotten by history. And many women used their mathematical or scientific abilities at home or in the fields or in business, not in university study and research, and so history was never interested in them. And many other women have never had the chance or the time to learn.

The women recorded in the history of mathematics are the tip of the iceberg. They are the proof that women can do great things in science and mathematics, but they also represent the mass of invisible women who were ignored, forgotten or deprived. We don't all have to be extraordinary women—any more than all men have to be

*This chapter first appeared in the 'Women in Mathematics and Science Kit' produced by the Participation and Equity Program, Victoria, Australia 1982. Reproduced with permission.

extraordinary because of Gauss or Einstein. But the extraordinary women give us the go ahead. They tell us that maths and science are part of a woman's world, just like everything else.

This is a history of women in mathematics. At least, it is a history of some women and some mathematics, in the Western world and in recorded history. A full history can never be written, because so many of the facts have gone for good. But we can learn some more about women and mathematics from the facts that are left to us.

We don't know how mathematics began. By the late Stone Age people were using systems of numbers. By 3000 BC people were putting up large stone buildings and crossing the sea in ships, so they must have known something about mathematics. However, we don't know how they worked all this out.

And we can only guess at the part women played. In ancient Egypt families played mathematical games together, and women ran businesses, which sounds as if women knew something about mathematics. But at the same time most Egyptian women were not taught to read or write, so the ancient Egyptians obviously did not think it was important to educate girls. We only stop guessing when we come to the ancient Greeks. From the sixth to the fourth century BC, the Greeks thought and studied and wrote and taught and changed the ways of thinking in the Western world. They took the ideas and knowledge of the older Babylonian and Egyptian cultures, and improved them and put them into a system.

The Babylonians, for example, had used mathematical formulas for surveying and business. But a Greek, Thales, was the first person to insist that mathematical formulas had to be proved. His pupil Pythagoras, went on bringing logic into mathematics. In 539 BC he started a colony in Corona. At the school there, he began the science of mathematics, or mathematics for its own sake.

Pythagoras was called 'the feminist philosopher'. Many Greek women had become famous for their thinking and writing, but in general Greek women were kept out of education and public life. However, Pythagoras wanted women to be part of his school, and at least twenty-eight women worked there, as students and teachers. Pythagoras's wife, Theano, wrote books on mathematics, physics, medicine and child psychology, and she and two of their daughters ran the school after Pythagoras died.

Another important school in ancient Greece was Plato's Academy, started in 387 BC. Plato believed that everyone should learn about everything. In his book *The Republic* he wrote, 'All the pursuits of men are the pursuits of women.' As a result women came to the

Academy in large numbers, although there was a law against women going to public meetings at the time.

Both Pythagoras and Plato thought that mathematics was very important, and even after the Christian era began, women went on studying mathematics. In a book written around the year 200, the Greek writer Athenaeus named several famous women mathematicians, though he did not describe their work. And nearly 200 years later the mathematician Hypatia was carrying on the Greek tradition in Alexandria, until she was killed because of her studies.

However, for a thousand years after the fall of Rome in 476, there were no new ideas in mathematics or science in the western world. Then, towards the end of the Middle Ages, Copernicus, Kepler and Galileo started to work out theories about the movement of the planets. Finally in the seventeenth century Isaac Newton brought all these ideas together in his *Principia*. The new theories needed a new kind of mathematics. Maria Agnesi, the physicist Laura Bassi and Emilie du Châtelet all worked on developments of Newton's ideas.

During the Middle Ages people had turned against education for girls. They did not even want girls to read and write, because the girls might be tempted to sin—though the nun Hroswitha remarked in the tenth century that knowledge was not dangerous, only the bad use of it. Hroswitha was one of the few women scholars of this time (she included some mathematical puzzles in her writings), because nunneries were the only places where women could learn and teach.

Women had had some freedom under the Roman Empire, and Italian women kept some of that freedom. Then, when the Turks captured Constantinople (Istanbul) in 1453, scholars fled from Constantinople to Italy, and started teaching there. A revival of learning began, which we call the Renaissance. Many women became famous as writers and scholars, scientists and mathematicians during the Renaissance and afterwards.

Maria Agnesi is the most famous of the Italian women mathematicians. However, though most of the women in this chapter are the only famous women mathematicians of the time, Maria Agnesi is one among a number of Italian women mathematicians. Talking about Clara Borromeo of Genoa, people said, 'No problem in mathematics and mechanics seemed beyond her comprehension.' And Diamente Medaglia wrote a book on the importance of women studying mathematics, where she said, 'To mathematics, let women devote attention for mental discipline.'

Italy was not paradise. At the same time as Maria Agnesi was learning half a dozen languages, other upper-class Italian girls were not even taught to read. But at least the extraordinary women were not frowned on in Italy. Emilie du Châtelet was unusual among French girls in being given a good education, and unusual among French women in admitting that she liked study.

There were good reasons for this. The seventeenth century playwright Molière had written two popular comedies where he sent up educated women, or 'blue stocking'. So '*femme savante*' (learned woman) became a term of criticism, and women tried hard to hide their brains. Most of the French thinkers of the time agreed with Molière. Jean-Jacques Rousseau wrote a book about new ideas in education, *Emile*, but his new education was for boys—he still wanted girls to be educated to be wives. For two hundred years the only famous French women were famous for their marriages, affaires and social skills.

Emilie du Châtelet set a good example as a woman who continued to study, and who wrote and talked about her studies. Many people criticised her, but she forced many others to admire her. However, even after the French Revolution where women played an important part, Sophie Germain used a male pen-name in writing to the mathematician Carl Gauss—'To escape the ridicule attached to a learned woman (femme savante)'—as she said. Molière's send up of the '*femmes savantes*' was still having its effect.

Italy encouraged the extraordinary woman; France frowned on her; and England, towards the end of the seventeenth century, began to approve of education for upper-class women, because educated men preferred educated wives. Mathematics was a new study for men at this time, and so women were allowed to join in. The *Ladies' Diary*, a popular magazine published once a year from 1704 to 1841, contained a section of mathematical puzzles. The magazine sold up to seven thousand copies a year, and a number of women sent in their own puzzles to it.

Mary Somerville, who later became a well-known writer of popular science books, was looking at a ladies' fashion magazine when she discovered 'what appeared to me to be simply an arithmetical question; but on turning the page I was surprised to see strange looking lines mixed with letters, chiefly X's and Y's, and asked; what is that? Oh, said Miss Ogilvie, it is a kind of arithmetic: they call it Algebra; but I can tell you nothing about it.' This story seems strange to us now. In the twentieth century women's magazines do not include problems in algebra, and no one would try to publish a magazine for women containing mathematics. But at the time of the *Ladies' Diary* mathematics was a new study, and most people had taught themselves mathematics. As mathematics became more complex, women were left out again. Men were taught mathematics, to train them for jobs in the new technological society: women's jobs were in the home.

Every time we see a new interest in mathematics—in ancient Greece, after the Renaissance, after Newton's work—we also see one or more famous women mathematicians. However, each time the woman mathematician is seen as unusual. Hypatia, Emilie du Châtelet, Maria Agnesi, Sophie Germain, Mary Somerville and Ada Lovelace were all odd ones out in their time. Their achievements did not lead to the education of other women.

In the nineteenth century women started thinking, talking and writing about education, as part of the first wave of the women's movement which looked at the whole inferior position of women in society. Women began to fight for the right to go to university, and with more chances to study there were more notable women mathematicians. Sonya Kovalevsky left Russia to study mathematics in Germany. Grace Chisholm Young decided to become a mathematician after her success at Cambridge. Emmy Noether, the greatest woman mathematician so far, planned to be a teacher, but went to the University of Erlangen because women had just been admitted.

But Sonya, Grace and Emmy still faced trouble in getting jobs. When Sonya Kovalevsky was finally given a job at the University of Stockholm, the Swedish writer Strindberg said, 'as decidedly as two and two make four, what a monstrosity is a woman who is a professor of mathematics.' Grace Chisholm Young married and worked at her mathematics by herself, while her husband taught and published in the public world, and Emmy Noether only had a job worthy of her talents for the last year and a half of her life. Once women were educated, they found more prejudices to overcome.

However, in the twentieth century more women were able to work as mathematicians. Hanna Neumann worked at the Australian National University until her death in 1971. Maria Pastori's parents could not afford to keep her at school, but she won a series of scholarships, and finally became a professor of mathematics at Milan. Another Italian woman, Maria Cibraio, is a professor of mathematics at Pavia, the second oldest university in Italy. In France, Jacqueline Lelong-Ferrand was one of the first girls to go to the famous Paris Ecole Normale Superieure of the rue d'Ulm: she is now professor of mathematics at the University of Paris. Sophie Picard was educated in Russia, but her family went to Switzerland after the Russian

Revolution. Sophie worked as an actuary after her father died, but she continued to study mathematics and finally became a professor at the University of Neuchatel.

All these women, and many more, show that good education makes good women mathematicians. However, many people still believe that girls can't do maths. Because they believe this, they make it come true. Girls are given less and worse teaching than boys, and so many girls come to believe that they really cannot do maths. Only the really keen women students continue with their studies in mathematics.

In the 1980s computers and machines are very important, and mathematics is very important in understanding computers and machines. While women are kept out of mathematics, women are kept out of a big part of our society. Certainly there are more women mathematicians now, but there are also a great deal more men mathematicians than in the past, because mathematics has become more important. So we still need to learn from the famous women mathematicians: girls can do maths, and, if they want, they can do it very well.

Index

'A' Levels, 4, 5, 169, 181
computer studies, 12
Achievement motivation, 82
fear of success, 82, 86
Agnesi, Maria, 78, 250–1
Arithmetic confused with maths, 40
Assessment of Performance Unit (APU), 5–7, 35–36, 38–9, 62, 64, 123, 157
Primary Survey 1, 33
Primary Survey 2, 33
Primary Survey 3, 23, 24, 32–3
Secondary Report 1, 74
Review 1979–1982, 35
Australia
survey of print media, 84–7
Avery Hill College Conference, 187–95

Ballerini, T., 150
Banking as girls' career, 53–5
'Be a Sumbody', 175, 187–92, 227
Bell, E.T., 9
Berlin, I., 227, 228, 236, 237
Books, maths, 110–21
illustrations, 112–14
infant, 111–14
junior, 114–6
male/female roles in, 114–6
Boys
constructive naughtiness, 67, 126
dominance, 242, 245–6
exclusion of girls, 150, 159, 242
household tasks, 104
Lego play, 160–1
manipulation of classroom, 174
parents' aspirations for, 94–100
public exam results, 4, 180
single-sex groups, 176–9, 181–4
in teacher/helper relationships, 130–2
see also Children's attitude to maths, Power in Class, Pupils' mathematical attainment, Sex-stereotyping, Teachers

Careers
advice, 51–7, 193
maths as prerequisite, 189–90, 225, 234
and technology, 247
CDT (Craft, Design and Technology), 226–7, 229, 232
Children's attitude to maths
at age 5, 11
confidence in own abilities, 42–3
perceived difficulty of, 42
emotions aroused, 40–1
at primary school, 22–5
sex differences, 45–7
as useful subject, 41–2
see also Boys, Girls, Pupils' mathematical attainment
Chipman, S.F., 2–3
Classroom practices, 124–45, 153–5
Cockroft Report, 14–15, 36, 188
Coercive inducements, 235, 238
Collaborative behaviour, 15, 183, 187, 245
Competitive behaviour, 15
Compulsory subjects, 236–7
Computation, 23, 25–7, 32, 35, 65, 70
Computers, 12, 241
girls' use of, 117–21, 150–1, 194, 242–7
owned by girls'/boys' families, 151, 241
schools' purchase of, 119–20
Computers in the Primary Curriculum, 150
Context-boundedness, 201–3
Critical thinking, 197
Croydon, 243–4, 246
CSE (Certificate of Secondary Education), 4, 5
Computer Studies, 12
Mode Three, 141
results, 5, 180, 233
Curriculum, 8, 12–14
DES Survey, 229
desirable topics, 34

Curriculum, (*Cont.*)
  primary favours boys? 31
  problem-solving, 14
  widening of, 205

Discovering Mathematics, 111
Du Châtelet, Emilie, 78, 250–1

Early Mathematics Experiences (EME), 111–13
Eddowes, M., 23, 195
Engineering as a career for girls, 52–3, 55–7, 230–1
Equal Opportunities Commission (EOC), 118, 121, 243

Fennema, E., 11, 68, 81
Fibonacci sequence, 219–20
Freedom, 224–39
  negative, 228–9, 233
  positive, 229–30, 236

GCE (General Certificate of Education) *see* 'A' Levels, 'O' Levels
Gender and Mathematics Association (GAMMA), 14, 22, 187
Gender roles *see* Sex stereotyping
Germain, Sophie, 9, 78, 251
Gilligan, Carol, 11, 199–206
Ginsburg, H.P., 70
Girls
  achievement motivation, 82–6
  and adult approval, 67
  constraint in course choice, 226–9, 235
  decline of maths achievement, 4, 154–5, 227, 233
    and compulsion, 237
    questioned, 61
  exclusion from boys' activities, 150, 159, 242
  femininity, 7, 144, 235–6
  good behaviour, 126–7
  household tasks, 104
  lack of confidence, 5, 242
  Lego play of, 160–2
  need of reassurance, 159–60
  public exam results, 4
  serialist learners, 74–5
  single sex groups, 176–9, 181–4
  social stereotyping, 8–12, 22, 82, 235–6

Girls, (*Cont.*)
  under-representation in maths courses, 3, 110, 153
  see also Children's attitudes to maths, Power in class, Pupils' mathematical attainment, Sex-stereotyping, Women
Girls into Science and Technology (GIST), 91–2, 96, 153

'Heinz dilemma', 198–202
  male/female responses to, 200–2
Herschel, Caroline, 78, 79
Hofstadter, D., 218
Holistic learners, 6, 68–9, 71–2
Homework, 95–6
Horizon (TV programme), 63, 193, 234
Hypatia, 78, 250

Individualised mathematics schemes, 139
  see also SMILE
Information technology (IT), 243–7
Information Technology in Schools (EOC), 243–4
Inner London Education Authority, 10
Interpretation of maths, male and female, 16

Kant, Lesley, 22
Kohlberg, L., 198–206
Kovalevsky, Sonya, 9, 78, 79, 252
Krutetskii, V.A., 69

Ladies' Diary, 80–1, 251–2
Language assumptions, 206–8
Language study, nineteenth century, 8, 59
Learning strategies, 6–7, 16, 68–75
Lego, 160–1
Lovelace, Ada, 78

Male-dominated jobs, 115, 241
Male-dominated subjects, 116, 225, 234, 238, 239
Marland, M. 13
Mathematics and the Ten-Year-Old (Schools Council), 62
Mathematics Counts (Cockroft Report), 14–15, 34, 188
Maths Adventure, 111, 113

Maths as high-status subject, 8, 9, 13, 58–60, 77
Maths as prerequisite subject, 189–90, 225, 234
Maths attainment, factors affecting, 123, 137
Measurement, 22–4, 27, 35
Microcomputers *see* Computers
Minority groups, 4
Moral education, 198–200
Morality, male/female interpretations, 201–2

National Child Development Study, 110
Noether, Emmy, 13, 78, 252
Nuffield *Mathematics – The First Three Years*, 111, 113

'O' Levels, 4–5
  and careers prospects, 53, 55
  in computer studies, 12
  importance of, 188
  mathematics candidates, 62–3
  results, 180
Oadby Beauchamp College, 163–85
Open University, conferences at, 175, 187
Osen, Lynn, 9

Parents
  aspirations for girls, 10–11, 94–5
  and children's educational qualifications, 94–5, 166
  and children's occupational aspirations, 98–110
  and children's sex-stereotyped roles, 93–5
  and children's toys, 94, 101
  expectations for girls, 78, 166
  helping with homework, 95–6, 166
  on importance of school subjects, 96–8
  middle class, 94, 101
  own sex-stereotyped roles, 11, 95–6, 102–3, 106, 118
  and pre-school child, 11
  school meetings, 95–6
  sexist views of, 101–3
  working class, 94, 101–2
Pask, G., 68–70, 72–5
People connectedness, 201–2, 206
Petersen, 11, 81
Plato, 249–50
Plowden Report, 67
Power in class, 125–42, 160
  girls in powerful position, 125–6, 128–9, 137–8, 141
  rule challenging, 126–7, 133–6, 142, 144–5
  subordination of girls, 125–31, 137, 159–60
Press as attitude-shaper, 83
Primary schools
  research project, 156–62
Prime numbers, 218–9
Procedures, 70
Problem posing, 209–20
  re-posing, 210, 219–20
Problem solving, 14–15, 26–7, 196–220
  dialogue in, 197
  ignoring of fundamental concepts, 211–12
  'situation to investigate', 209, 216–18
  students' inability to solve, 212–13
  and the two cultures, 196, 204
'Progressive' teaching, 71
Pupils' mathematical attainment, 10, 22–35, 143–5, 153–5, 180
  age 10, 25–31
  age 11, 32–3, 35, 48–9, 61, 64
  age 15, 48–9. 188
  at 'A' Level, 169–70
  serialist learners, 74–5
  sex differences, 47–50, 61
  see also Boys, Girls, Maths as high-status subject
Pythagoras, 249

Questionnaires
  'Be a Sumbody', 190–2
  Parents', 94–108
  Pupils', 170–4

Roles, masculine and feminine *see* Sex-stereotyping

Schools Council, 22
  Project, 24, 62, 65, 163
Self-confidence, 8
Serialist learners, 6, 68–75
Sex Discrimination Act 1975, 1, 231–3
Sex-differentiation in schools, 164–5
Sex-stereotyping, 10–11, 90–108, 193–4, 223–9
  and childrens' household tasks, 103–4
  in children's maths books, 10. 110–21
  and computer use, 117–21, 149–50

Sex-stereotyping, (*Cont.*)
and Craft, Design and Technology, 149, 226–7, 239
at home, 104
middle-class parents and, 94
and parents' educational aspirations, 93–8
and parents' occupational aspirations, 96, 98–100, 107
positive discrimination, 149–50
relations between home and school views, 90
and school subjects, 96–8, 107
school view, 91
voluntary activities of children, 105–6
working class parents and, 94, 101–2
see also Parents, Teachers
Shuard, H.B., 36
Single-sex groups, 176–9, 181–4, 187–8
SMILE (Secondary Maths Individualized Learning Experiment), 73, 131–5, 192
Social environment of maths success, 77–8, 137
cultural pressures, 81–2
Somerville, Mary, 78–9, 251
Spatial ability, 2–3, 23–4, 25, 28, 35

Tall Trees School, 92–3
parents' questionnaire, 94–108
Teachers
attitudes, 2, 7, 166, 167, 176–7
boys claiming attention of, 142, 153, 174–5, 246
and computers, 119–20, 149, 224
evaluation of children's achievements, 31–36
girls claiming attention of, 127–43, 153, 174–5
inspiring girls' efforts, 154–5
planning and executing research project, 157–75
poor mathematical qualifications of, 71–2
and public exams, 4
and sex-stereotyping, 91, 166

Teachers, (*Cont.*)
social conditioning and, 6, 7
strategies, 70–2, 75
teaching priorities, 31, 65–7
see also Power in class
Three-dimensional visualisation, 71
Timetabling as choice limiter, 226, 232, 242
Toys, 151, 157
Lego, 160–1
Transition from primary to secondary school, 122, 131

United States research, 1–4
differences in UK and US findings, 6, 122

Versatile learners, 69–82

Walden, R., 61–3, 74, 157, 175
Walkerdine, V., 61–3, 74, 157, 159
Ward, M., 25, 34, 36
Weighing, 35
Wilson, D.M., 2–3
Women
conflicting feminine roles, 10, 85, 86
eminent mathematicians, 8, 13, 78–81, 194, 248–53
fear of success, 82–3, 186
in maths books, 114–16
need to work harder, 85, 86
negative consequences of success, 82–3, 87
percentages in different types of job, 84–5
poor performance genetic? 2
roles as seen by press, 84–7
and traditionally male jobs, 51–6, 230–1, 241–2
see also Girls, Sex-stereotyping
Word problems, 14, 25, 33

Young, Grace C., 9, 252